Productivity PLUS+

Gulf Publishing Company
Book Division
Houston, London, Paris, Tokyo

Productivity PLUS +

How today's best run

companies are gaining

the competitive edge

John G. Belcher, Jr.
of the AMERICAN PRODUCTIVITY CENTER

Productivity Plus

Library of Congress Cataloging-in-Publication Data

Belcher, John G.
Productivity plus.

Includes index.
1. Industrial productivity.   2. Industrial
management.   I. Title.

HD56.B45 1987	658.3′14	87-9222

ISBN 0-87201-451-7

First Printing, October 1987
Second Printing, August 1988

Contents

IV: Continuous Productivity Improvement, 177

Acknowledgment

The American Productivity Center, a nonprofit organization located in Houston, Texas, has contributed to this book in many ways. I have learned a great deal about the subject of productivity through my association with the Center over several years, both as vice president and in an adjunct consultant relationship. The Center's excellent publications are widely referenced in the text, and its people have provided invaluable input. I am particularly appreciative of Carl Thor, Stu Winby, Diane Riggan, and John Younker for their specific contributions to this book.

Preface

The decline in this nation's productivity growth rate has been much lamented in recent years. The contribution of this decline to the economic woes of the United States has been well documented.

The American business community has responded to this plight by making productivity a major organizational issue. As a result, productivity improvement programs have proliferated and the nation's productivity growth rate has recently improved. Many of these programs, however, have failed to meet expectations. Even where initial results have been positive, they frequently fail the test of sustainability.

Productivity Plus suggests that productivity, because of its long-term strategic importance to the organization, should be approached as a management process rather than as a program. The philosophy underlying this approach has evolved over the last several years as the American Productivity Center has observed hundreds of organizations across a wide variety of industries and assisted them in improving productivity.

The intent of this book is to define and clarify a productivity management process and to raise and discuss the issues that must be addressed to ensure that productivity improvement will be comprehensive and continuous.

John G. Belcher, Jr.
American Productivity Center
Houston, Texas

To my wife, Nancy Newhard, whose patience and support make my work possible.

I

PRODUCTIVITY AND ITS MANAGEMENT

1

The Challenge of Productivity

WHAT IS PRODUCTIVITY?

Conceptually, productivity is a simple notion: It is the relationship between the output of an organization and its required inputs. We can quantify productivity by dividing the outputs by the inputs. We increase productivity by improving that output/input ratio; that is, by producing more output, or better output, with a given level of input resources.

Though it seems simple, misconceptions about productivity are widespread. *Productivity* is often equated with *production*, for example. If more output of goods and services is attained, then productivity is assumed to have increased. But production only represents the top half of the equation; we cannot reach a conclusion about productivity without considering the changes in inputs that were required to improve the output.

Another common misconception relates to the definition of *input*. Managers and nonmanagers alike often assume that the word productivity applies exclusively to the labor input. This assumption has no rational basis, as an organization's success is dependent upon the effectiveness with which it utilizes *all* of its resources—raw materials, capital equipment, and energy, as well as labor. The view that productivity improvement only applies to the labor input is a dangerously narrow one. It may well result in a failure to capitalize on significant opportunities to improve organizational performance through better equipment utilization, reduction in materials losses, and conservation of energy.

Beyond the common misconceptions, the productivity issue is fraught with practical complexities. An organization's output, for example, may be difficult to define. The output of a manufacturing organization may be obvious to the productivity observer, but what is the output of a bank or an engineering department or a government agency? Many nonmanufacturing organizations, unfortunately, have failed to deal effectively with productivity because of an inability to relate the term to their business. Indeed, an attempt to apply the traditional, manufacturing-oriented definition of productivity in a white-collar environment will likely fail to produce the desired results.

Another complicating factor is quality. Does the more effective utilization of resources while producing defective products or services qualify as productivity improvement? What exactly is the relationship between quality and productivity?

This list of complexities could continue at great length. The apparent simplicity of the output-over-input concept is deceiving.

HISTORICAL PERSPECTIVE

Prior to the mid-1960s, labor productivity growth in the United States was material and consistent. From the end of World War II until 1965, national labor productivity advanced at an average annual rate of 3.2%. A slowdown in the growth rate became apparent after 1965 and worsened during the 1970s, with average improvements barely exceeding one percent during the middle years of the decade. The trend appeared to bottom out in the 1978–1982 period, when labor productivity actually declined by an average of 0.2% per year. We then experienced a comeback, with productivity increasing at an average annual rate of 2.5% from 1982 to 1984. While some observers declared that our productivity problem had been solved, subsequent events proved them wrong—national labor productivity increased by only an average of 0.6% in 1985 and 1986 (see Figure 1-1).

When we factor capital into the productivity growth equation, results are even more disturbing. *Total-factor productivity* (the relationship between

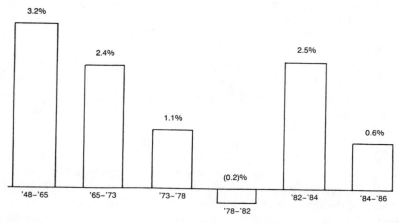

Figure 1-1. U.S. labor productivity growth—average annual change (non-farm business). Source: American Productivity Center.

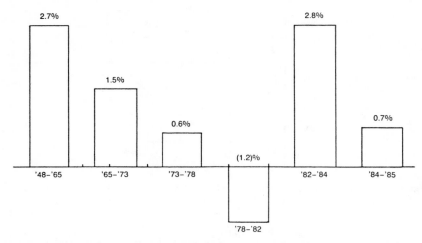

Figure 1-2. Total factor productivity growth—average annual change (business economy). Source: American Productivity Center.

national output and the sum of the tangible capital and labor inputs) mirrors that of labor productivity alone, and even trends lower in most years (Figure 1-2).

How does U.S. performance compare with that of other industrialized nations? Not favorably; our labor productivity growth rate since the mid-1960s has been exceeded by virtually every other industrialized country in the world (Figure 1-3). We may find solace in the fact that, in absolute terms, we are still the most productive country in the world (Figure 1-4). But national data can be misleading. A landmark study by the American Productivity Center indicated that a comparison of productivity levels be-

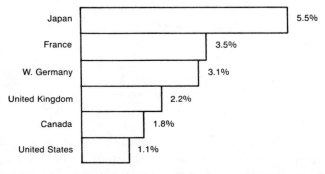

Figure 1-3. International labor productivity growth—average annual change, GDP/employee. 1960–1986. Source Bureau of Labor Statistics.

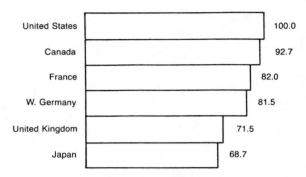

United States	100.0
Canada	92.7
France	82.0
W. Germany	81.5
United Kingdom	71.5
Japan	68.7

Figure 1-4. International labor productivity levels, GDP/employee. 1986. (U.S. = 100.) Source Bureau of Labor Statistics.

Table 1-1
Labor Productivity Level, Japan vs. U.S., 1983
(U.S. = 100 for each industry, 1975 Exchange Rates)*

Private Business Economy		68
Goods **(69)**	Agriculture	36
	Mining	69
	Construction	53
	Manufacturing	88
	Food & Tobacco	53
	Textile	41
	Pulp & Papers	81
	Chemicals	156
	Primary Metal	132
	Fabricated Metal	48
	Non-Electrical Machinery	89
	Electric Machinery	173
	Transportation Equipment	105
	Other Manufacturing	65
Services **(64)**	Transport & Comm.	41
	Utilities	87
	Trade	55
	Finance & Insurance	178
	Business & Professional	50

* Source: American Productivity Center.

tween the United States and Japan would yield dramatically different conclusions depending upon the industry evaluated. As Table 1-1 shows, the Japanese are significantly more productive than their U.S. counterparts in certain industries.[1]

In any event, given the present differentials in international growth rates, the United States may soon forfeit its world leadership in productivity. This persistent erosion of our productivity advantage has enormous implications for our nation's ability to compete in world markets.

The causes of our present predicament have been thoroughly analyzed and debated in the media. Suffice it to say, the villains are many and varied; some of the causes commonly cited by economists include declining capital investment, increased government regulation, the changing industrial mix of our economy, declining expenditures for research and development, and changes in the composition of the work force. Some would suggest that the most significant cause, the effect of which cannot be quantified by the economists, is the complacency of American managers engendered by the unprecedented economic growth and this country's dominance of world markets during the fifties and sixties.

IMPLICATIONS TO THE FIRM

Apart from its implications for our national economic well-being, what is the significance of productivity to the business enterprise? A look at some comparative data for the United States and Japan during a recent ten-year period (Table 1-2) provides some insight into the answer.

Table 1-2
Average Annual Changes in Compensation, Productivity, and Unit Labor Costs
1973–1984*

	Compensation	Labor Productivity	Unit Labor Costs
United States	8.8%	2.0%	6.7%
Japan	8.1	7.3	0.8

* Source: Bureau of Labor Statistics

During this period, compensation costs in the United States and Japan increased at about the same rate, 8.8% vs. 8.1%. Changes in labor productivity were dramatically different, however (2.0% vs. 7.3%), and were mirrored in unit labor costs (the amount of labor cost contained in a unit of product or service).

A similar analysis could be made for any input factor—labor, capital, materials, or energy. The amount of cost contained in a unit of product or service is basically dependent on two things: the prices paid for the various inputs and the effectiveness with which those inputs were utilized in the production or service delivery process. Assuming that competing companies pay about the same prices for their inputs (not necessarily a good assumption in international markets, but probably reasonably true for domestic competitors), the inescapable conclusion is that the company that is most productive will have the lowest-cost product or service. The data in Table 1-2 reflect this fact.

The role of productivity becomes even more prominent for those organizations facing international competition. Wage levels in many countries are substantially lower than those in the United States. Under these circumstances, the productivity of American companies must be commensurately greater than their foreign competitors to maintain parity in product costs.

In evaluating the impact of productivity on the organization, consider the following equation:

$$\Delta \text{Costs} - \Delta \text{Productivity} = \Delta \text{Prices}$$

If a business organization is to maintain its level of profitability from one period to the next, this mathematical relationship must prevail. The organization must pass through to its customers, in the form of higher prices, the net effect of increases in unit input costs offset by increases in the productivity of its inputs. If the unit costs of an organization's resources, for example, increase 6% while productivity increases 4%, the selling prices of the company's products and services need rise by only 2% in order to maintain current levels of profitability.

This suggests that productivity is the only weapon that an organization has, other than price increases, to offset the effects of cost increases on the bottom line. An increase in productivity results in a commensurate reduction in the pricing required in order to offset cost increases and maintain profitability—a phenomenon of immense significance in a competitive environment.

On the other hand, if productivity declines (a common occurrence in American organizations in recent years), the equation dictates that selling prices be increased proportionally more than is required to offset input cost increases. Such a scenario is obviously not a tenable one in the long run for most organizations.

Upon further reflection, one can logically conclude that changes in profitability result, in a broad sense, from just two factors: productivity (the effectiveness with which we utilize all of our resources) and price recovery (the degree to which increases in unit costs of inputs are recovered by increases in selling prices). An organization that increases its profits has either utilized its resources more effectively (improved productivity) or has raised its selling prices to such a degree that incurred cost increases have been over-recovered, or both. Unfortunately, many business organizations lack the capability of analyzing profitability in this context—a critical shortcoming, as a simple illustration will demonstrate.

Suppose that two competing companies (Company A and Company B) have both realized significant, and nearly identical, profit growth in recent years. A cursory review of their financial statements might lead the observer to conclude that both organizations were equally healthy. When these companies' profit growth are viewed in terms of productivity and price recovery, however, a different picture emerges (Figure 1-5).

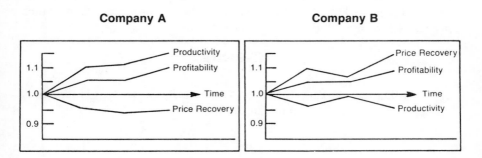

Figure 1-5. Company A versus Company B—profitability, productivity, and price recovery.

Company A's profit growth has resulted from productivity improvement; in fact, the magnitude of the productivity increase has been sufficient to more than offset declining price recovery. In other words, Company A has not fully passed through to the consumer its cost increases, but has, nonetheless, realized rising profitability through gains in productivity.

Company B, on the other hand, has achieved its profit growth in a different fashion. Productivity has declined at Company B, yet profit growth has been realized through rising price recovery. Company B has raised its selling prices more than is justified by increases in the unit costs of its inputs in order to offset declining productivity.

Now that we have analyzed the growth in profitability at the two companies, can we still conclude that they are equally healthy? Emphatically, no. While Company B, under certain conditions, may successfully pursue the described course of action in the short term, it must ultimately become uncompetitive vis-a-vis Company A. In a competitive environment, a Company-B strategy simply is not viable in the long run; if Company B does not reverse its trend of declining productivity, it will lose business and will not survive. Unfortunately for Company B, its information and analytical systems are probably not sufficiently developed for it to determine the relative impact of productivity and price recovery on its bottom line. Rising profits have lulled the management of Company B into complacency and have masked a potentially fatal illness.

Should productivity be a concern only to those organizations facing the rigors of competition? In organizations with protected markets, such as utilities, full rate relief (price recovery) may be impossible in these days of volatile energy costs and unsympathetic public utility commissions. In such an environment, productivity is a virtual necessity if stockholder returns are to be maintained. And for government, productivity improvement is the one practical alternative to higher taxes or reduced services, both of which are politically undesirable. It is difficult to imagine any organization for which productivity is not vitally important for success.

REFERENCE

1. "New Data Compares U.S. and Japanese Productivity," *Productivity Letter*. Houston, TX: American Productivity Center, August, 1986.

2

The Productivity Improvement Process

A common response to the productivity problem can be illustrated by reviewing the experience of the Consolidated Software and Sprocket Company (a conglomerate, obviously).

Consolidated's chief executive officer had reason to be concerned about productivity. The company's Sprocket Group was being battered by Japanese competition and was losing money. The Software Group, on the other hand, had been slow to respond to rapid technological changes and was not meeting the CEO's expectations of rapid profit growth.

Being a decisive executive, the CEO called together his group vice presidents and directed them to improve productivity in their respective organizations. He stressed that he didn't care how they went about it, as long as results were achieved. He had, however, recently read a great deal about quality circles and did feel that both organizations could benefit from more employee participation. In any event, he expected quick improvement, and although he personally would be occupied with acquisitions and other strategic pursuits, he expected regular reports on the efforts.

The vice president for the Sprocket Group was not anxious to launch a formal productivity improvement effort at this time. Substantial numbers of his hourly employees had been laid off, and reductions in the salaried work force were forthcoming. Labor relations in his heavily unionized operations were at a low point. In addition, he had recently begun a crash development effort to upgrade the group's products in order to meet the Japanese challenge, and he did not want to distract his management's attention from this critical project.

Nonetheless, the CEO's mandate was clear and could not be ignored. The vice president called together the Sprocket Group's management team and

11

announced that productivity was the organization's new priority. Sales per employee was to be the group's productivity measure, and all managers were expected to contribute to improvements in that measure over the next six months. Unfortunately, the product development program would have to be scaled back, as this effort was not expected to yield incremental sales for 9–12 months, and current manpower expenditures would adversely affect the sales-per-employee measure.

Since it was clear that the CEO had become interested in participative management, a consultant was retained to install quality circles in the Sprocket Group.

The Sprocket Group consisted of several divisions, and the specific approaches to productivity varied considerably. One division manager concentrated on technical improvements, focusing his engineers' attention on methods and machine cycle times. Another sought improvements through the application of various motivational techniques. A third manager simply took further reductions in his work force. Still another manager instituted marketing promotions and special discounts in an effort to boost the sales side of the equation.

After six months, it was clear that the Sprocket Group's productivity improvement efforts were going nowhere. The disaffected workforce resisted the effort because of fears of further job loss. Middle managers and first-line supervisors largely ignored the initiatives, assuming that senior management's interest in productivity would soon pass and things would return to normal. Because there was no overall coordination or guidance, results from division to division were highly variable, ranging from modest improvement to deterioration. In some divisions, the imposition of a productivity improvement program at a time when managers were preoccupied with serious organizational and marketplace problems resulted in disruption and disarray. And the reduction in the product development effort caused further loss of market share to the Japanese.

The Sprocket Group's quality circle program was a dismal failure. Since management had shown no previous inclination toward involving hourly workers, employees were incredulous and suspicious of management's motivations. The union, which had not been consulted or involved in planning for the quality circle program, actively resisted the effort. After six months, several circles were going through the motions, but no measurable improvements were discernible.

It appeared to the CEO that after six months of effort, the Sprocket Group was in worse condition than before the productivity program was initiated. The group vice president lost his job.

The productivity effort proceeded somewhat differently for the Software Group. The group vice president concluded that the productivity program should be publicized and made highly visible to employees. He also was convinced of the need for a coordinating entity; while the Software Group was not as large and diverse as the Sprocket Group, it was, nonetheless, a complex operation.

He had the perfect candidate for productivity coordinator—a bright young MBA presently in the Software Group's management training program. This individual was relatively inexperienced but had demonstrated considerable promise. What better way to develop him than to appoint him productivity coordinator for the Software Group. Since the group vice president didn't have time to supervise him personally, he would have the coordinator report to the employee relations manager.

The productivity coordinator mounted an impressive publicity campaign. Posters went up throughout the organization. A catchy slogan was adopted. The group newsletter exhorted employees to greater heights of productivity. The coordinator also began meeting with each of the group's managers in order to provide guidance to their improvement efforts.

Like his counterpart in the Sprocket Group, the Software Group vice president was sensitive to the CEO's suggestions, and he too installed quality circles. Conditions in the Software Group were more conducive to the quality circle approach: employee relations were good and the organizational style was relatively open and participative. Early circle results were encouraging.

After six months, sales per employee in the Software Group had increased three percent, and naturally, the CEO was pleased. The group vice president was named to replace his terminated counterpart in the Sprocket Group to perform the same magic there.

Unfortunately, while short-term results were good, productivity improvement could not be sustained in the Software Group. The publicity and hoopla generated some initial interest and activity, but employees recognized the effort for what it was: a "program." It was perceived as being no different from last year's product development program or the quality program of the year before. As such, the initial enthusiasm and attention waned, the posters grew stale, and employees awaited the call to action for the next program. Quality circles, after a promising start, gradually lost their effectiveness and were ultimately discontinued.

The work force, which consisted largely of technical and professional people, also had difficulty relating productivity to their work. The word carried an "efficiency" connotation, and employees questioned whether that was an appropriate focus for their improvement efforts.

The productivity coordinator became the most frustrated employee in the Software Group. His efforts were heroic, but as a new employee occupying a relatively low position on the organization chart, he had little influence with the more senior members of the organization. He ultimately resigned to join one of Consolidated's competitors.

THE PITFALLS

The experience at Consolidated illustrates a number of pitfalls that are common to many productivity improvement efforts. Those described below occur with predictable regularity.

Mandate from top management. The senior executive of a company or organizational unit will often call together his subordinates and announce that productivity improvement is required. The typical result is a scramble by the subordinates to achieve some quick gains (which they often do through brute force), but the ultimate result is an uncoordinated, nonsystematic, and ill-informed effort that simply ensures that gains will be sporadic and unsustainable.

The president of a major energy concern ordered the company's executives to implement a major cultural change in order to increase productivity and performance. He felt strongly that a more positive climate, characterized by teamwork and involvement, was necessary if significant performance gains were to be realized. He made it clear that he expected quick results, and he did, indeed, get action. Frantic meetings were held, consultants were called, and the mandate worked its way down the hierarchy. Enormous amounts of energy were expended by managers trying to figure out what they were supposed to do and how they were supposed to do it. The result? Three years later, the mandate was forgotten and the company's culture was not discernibly different.

Unclear definition and rationale. Employees in the typical organization lack a clear understanding of productivity and its implications. Productivity improvement to them simply means "working harder," and the outcome is perceived to be loss of jobs. The company and its managers are seen to be the sole beneficiaries of productivity improvement; there is nothing in it for workers. Where these perceptions exist and persist, it is impossible to enlist employees' cooperation in the improvement effort. The work force resists, and the company is working against itself.

Weak commitment from the top. Senior executives are quick to extol the benefits of productivity improvement and to profess their support for productivity improvement efforts, but their actions often belie their words. They are not personally involved in the effort, they do not provide resources to support it, and they fail to modify existing management systems to support and reinforce productivity improvement on an on-going basis. The result is business as usual.

The appoint-a-coordinator trap. It will be suggested in the next chapter that a coordinating entity is an important element of the productivity management process. Ill-conceived decisions on this issue, however, can be very damaging. Executives often do not position this person properly and fail to provide adequate resources to support his efforts. The individual selected often has little influence with the rest of the organization and is unable to bring about the necessary changes. The improper positioning and selection of the coordinator sends a very negative signal to the organization at large; it suggests that productivity really is not as important as had been indicated.

One service organization created such a job as an entry-level position and staffed it with a very young, newly-hired individual. The person was capable, but lasted less than a year, as the task was impossible. An aerospace company, on the other hand, created a highly-placed position but filled it with a long-term employee who was ineffective in his previous position, in part because of his inability to get along with others in the organization. His influence with other managers was, of course, negligible, and his success was limited.

Failure to assess organizational readiness. Companies often undertake a full-scale productivity improvement effort without consideration of their readiness to do so. A number of important questions simply are not asked. For example, are relationships with employees or the union so strained that the effort will meet heavy resistance? Is organizational awareness of the implications of productivity so low that employees will misunderstand the purpose and rationale for the effort? Is the management style of the company so autocratic that managers and supervisors will be unable to effectively involve employees in the effort? There may be a number of organizational barriers that, if not addressed as a prelude to the improvement effort, will ultimately ensure its failure.

The measurement hang-up. While productivity measurement is an important reinforcing element of an effective productivity management process, it too often becomes both a prerequisite and an end in itself; the numbers drive the organization, causing counter-productive behaviors. This pitfall is evident in the often-heard statement, "There is no point in our trying to improve productivity until we can measure it."

One company whose executives subscribed to that philosophy spent two years attempting to develop elaborate productivity measurement systems as a prerequisite to their improvement effort. Unfortunately, they lost competitive ground during that period because they did not attend to productivity *improvement*. Measurement is indeed a valuable tool, especially when used in a supportive and reinforcing fashion, but we have yet to discover a physical law that precludes improving productivity without precise measurement systems.

Unclear responsibilities and weak accountabilities. When managers are asked who is responsible for productivity improvement, they invariably answer, "We all are." While no manager would have the temerity to suggest that he is not responsible for productivity, that responsibility is typically ambiguous and implied, rather than explicit. In reality, managers spend little time working on productivity or on developing an environment in which subordinates contribute significantly to productivity improvement. Until responsibilities are explicit and unambiguous, productivity improvement cannot become a driving force in the organization.

Fascination with techniques. There are innumerable techniques designed to bring about productivity improvement—quality circles, value analysis, suggestion systems, work measurement, office automation, to name a few. Organizations tend to become enamored of techniques. The sequence of events is predictable: A certain approach begins to receive favorable press, success stories abound, and managers everywhere rush to implement that technique in their organizations. Little thought is given to whether that particular technique is the most effective one, given their organizational circumstances and situation. This mania is exemplified by quality circles, which now exist by the thousands, often in organizations that are not ready for them, are not able to capitalize on the benefits they offer, or have lost sight of their intended purpose.

This pitfall has another, more serious interpretation. The typical productivity improvement effort makes liberal use of improvement techniques, but does not impact the environment within which productivity improvement must occur. Productivity improvement is not reinforced as a basic and integral element of organizational functioning, and organizational systems, practices, and procedures are not modified to support and reinforce productivity improvement on a day-to-day basis. When the selected techniques have run their course and exhausted their potential, the entire effort becomes moribund.

PRODUCTIVITY AS A STRATEGIC ISSUE

Most of these pitfalls are symptomatic of a deeper problem: the tendency to lose sight of the strategic implications of long-term productivity improvement and to approach productivity as a "program." A program, by its very nature, carries the seeds of its own destruction. Employees assume that the productivity program, like all other programs before it, will soon pass. As such, the typical productivity program may yield some short-term gains, but fails to promote continuing and meaningful improvement.

A full comprehension of the implications of productivity (as discussed earlier, productivity improvement is ultimately essential to survival in a competitive environment) demands that productivity be addressed as a strategic issue by the organization. Unfortunately, while many companies devote considerable management time and effort to strategic planning, addressing such issues as market share, product development, and technology, relatively few explicitly establish productivity improvement as a strategic objective. This is a critical oversight; for many organizations, the implications of productivity improvement to the future health and prosperity of the organization equals or exceeds those of the more traditional strategic issues. Examples are not hard to find—the American automobile and basic steel industries have fallen from positions of market dominance in large part because of their inability to keep pace with the rising productivity of international competitors.

While productivity merits treatment as a strategic objective, it is also related to other strategic objectives of the organization. If growth in market share is an accepted corporate objective, for example, is not productivity improvement a major means of achieving that objective? Too often, the link is not explicitly made.

Even those organizations that do recognize productivity as a strategic issue often lack well-formulated plans for managing productivity to achieve continuing, long-term improvement. Generally, management envisions productivity improvement as a series of annual "cost reduction programs" that consist of a fixed number of defined projects. Can management really believe that it is maximizing the company's productivity improvement potential when its efforts consist of 12 technical projects managed by the engineers? Does the other 90% of the organization have nothing to contribute?

A PROGRAM OR A MANAGEMENT PROCESS?

Given the recognition of productivity as a strategic issue to the organization, a productivity "program" is clearly an inadequate response to the challenge. But what is the alternative? The alternative is a *management process* for productivity improvement, which contrasts with a "program" in several significant ways.

A program, by definition, is limited in its time frame. It is of finite duration and has a beginning and an end. A management process, by contrast, is *continuous*. Once established, it is an on-going and integral element of organizational functioning.

Productivity programs are typically tactical in nature. As suggested earlier, programs often consist of one or more tactics or techniques designed to bring about productivity improvement at the working level. While these techniques are important elements of a management process, they are inadequate in and of themselves. A management process has a more *strategic* orientation; with productivity improvement recognized as critical to organizational success, the focus is on creating a climate and culture where productivity improvement is a way of life and is a part of everyone's day-to-day responsibilities. Within such an environment, the tactics and techniques utilized to bring about improvements have a meaningful context and are much more powerful.

A program, almost by definition, is isolated from the mainstream activities of the organization. It is an overlay or add-on activity and, as such, is not part of the continuing day-to-day activities of the organization and its employees. A management process, by contrast, is *integrated* into the other management systems and other organizational processes and practices. It is an integral element of the reward system, the budgeting and planning systems, the goal-setting practices, the communication systems, and the human resource management practices. Through this integration, productivity improvement ultimately becomes ingrained into the organizational culture and takes on a life of its own.

Productivity programs are typically narrow in scope—that is, they may address only labor productivity, or perhaps even just hourly or clerical worker productivity. A management process, on the other hand, is *comprehensive*, in that it addresses the productivity of all the major inputs to the productive process. The productivity of the organization's professional, technical, and management employees is pursued vigorously, as is the productivity of capital, material, and energy resources.

Finally, productivity programs are generally top-down in nature—that is, they involve only management people, with the bulk of the work force having no personal involvement in the effort. The result is lack of ownership and resistance on the part of the work force. A true management process, by contrast, *involves* everyone in the organization. If productivity improvement is to become an integral element of organizational functioning, all employees must be fully aware of its implications, must be committed to its improvement, and must have opportunities to contribute to the improvement process.

We might envision a continuum of productivity improvement efforts, from the pure program approach, which is clearly tactical and short-term in nature, to the management process approach, which constitutes a long-term effort to change the nature of the organization so that productivity improvement becomes institutionalized as part of the culture, integral to the day-to-day functioning of the organization.

Managers and employees must view productivity as an explicit and critical responsibility of their jobs. When employees perceive productivity as an add-on responsibility, something that is secondary in importance to the execution of their regular day-to-day activities, then productivity has not advanced beyond "program" status.

The manifestations of an environment into which productivity has been integrated are many. Productivity measures are not reviewed separately from financial results by senior management, but are an integral part of the process of analyzing and interpreting financial data. Productivity goals exist at every level, have wide visibility, and are closely monitored. Productivity is a regular agenda item in management meetings. Productivity is a key element of the performance evaluation process, and the reward system reinforces productivity improvement. All employees understand the implications of productivity and are actively involved in identifying improvement opportunities. Top management involvement and commitment to productivity is clearly and constantly visible to the entire organization. Productivity improvement is institutionalized as an operating norm, an ongoing and self-sustaining process.

Institutionalizing productivity is a long-term undertaking. It is not susceptible to a "quick fix." It is, indeed, an organizational change effort.

3

Top Management Commitment

WHY IS COMMITMENT SO VITAL?

Many, if not most, efforts to create a productivity management process never really get off the ground. Managers pay lip service to the effort but resent the distraction. Employees withhold their support out of mistrust of management and fear of losing their jobs. Organizational systems for planning, rewarding, communicating, controlling, and evaluating continue to function as they have for years or decades. Gains are elusive and not sustainable. The effort never advances beyond "program" status and ultimately dies a slow death.

The usual reason for this grim scenario is a lack of senior management commitment to change. Establishing a productivity management process is a change effort, and change never comes easy. You cannot mandate change, and you cannot delegate responsibility for change; attitudes and behaviors are too ingrained, and organizational systems too entrenched. Senior management must not only raise the banner, but must also lead the charge.

An anecdote may be instructive. The following conversation (as close as can be recollected) actually took place, in the author's presence, between the chief executive officer of a holding company and the president of one of the company's subsidiaries:

CEO: "Well, how is your productivity program going?"
President: "Not too well. I'm having difficulty convincing my people that this initiative is really important to the company."
CEO (with eyebrows raised): "Why is that?"
President: "Well, you seem to be interested only in acquisitions. That's where you spend all your time, and that's your reputation on Wall Street. You seem to have no interest in improving existing operations, so people think it's not important."
CEO (voice rising): "Wait a minute! At our management meeting six months ago, I told you and your people that I wanted to see operations improvement."

President: "But have you so much as inquired about our efforts since then?"

CEO (visibly angry): "Why should I have to ask about it? I said it was important. Isn't that enough?"

Obviously, it was not enough. The CEO pronounced his interest in productivity and then returned to pursuits that were more worthy of his time and energies. The signal that he sent to the organization was powerful; his actions belied his words. Is it any wonder, then, that the subsidiary's staff yawned when their president reminded them of the CEO's call for productivity improvement?

To successfully achieve the change implied in developing a productivity management process, senior management must not only be committed, but must be *involved*. They must prove their commitment to the organization.

TOP MANAGEMENT CONSENSUS

As a prelude to commitment and involvement, it is imperative that the organization's senior management reach a consensus on several issues:

☐ The definition of productivity, as it relates to their organization
☐ The strategic implications of productivity improvement
☐ The priority to be accorded to productivity improvement
☐ The need to devote resources to productivity improvement
☐ The role of the human resource in productivity improvement
☐ The nature of the improvement initiative—an organizational change effort or a program

Unanimity of opinion about productivity, even among top management, is generally absent in the typical organization. The lack of a clear consensus about productivity, and its relationship to other organizational goals, will limit the potential of the initiative before it gets off the ground. In such an environment, the degree of management commitment and attention will be variable, at best, and the effort will suffer accordingly.

The question of definition merits special treatment, as this issue often represents a major stumbling block. This is particularly true in service organizations and in companies with a heavy complement of technical and professional employees. If top management does not provide a meaningful definition of productivity, employees will interpret the term in a traditional fashion, i.e., "producing more widgets with fewer people," and will not accept its relevance to their functions.

During a computer network conference (a conference involving widely dispersed people connected via computers) sponsored by the American Productivity Center, the question of definition was addressed by productivity

II

COMMITMENT AND ORGANIZATION

managers in a variety of industries. The participants shared the definitions adopted by their companies, and some examples follow:

☐ Minimizing the amount of resources used in producing or delivering products or services of desired quantity and quality in a timely manner
☐ The quality, timeliness, and cost effectiveness with which an organization achieves its mission
☐ Better use of controllable resources to produce goods and services
☐ The change over time in the ratio of the volume of outputs (products and/ or services) produced and the resources used in producing those outputs
☐ Doing the right things right the first time
☐ People accomplishing something required (effectiveness) while using minimum resources (efficiency)
☐ Making changes to ensure the optimum effectiveness of an organization

There is no "right" definition; it is incumbent upon senior management to articulate a definition of productivity that people at all levels and in all functions can relate to.

THE PRODUCTIVITY DIRECTION STATEMENT

Once consensus has been achieved, management should become concerned with communicating the results of their deliberations to the organization at large. This must, of course, be accomplished eventually, for how can productivity improvement become an operating norm without a common understanding of top management's position on the subject?

Many companies have chosen as the communications vehicle the productivity philosophy or direction statement. This statement becomes an official document and carries the force of policy. It serves not only to communicate senior management's views on productivity, but also provides direction to decision-makers at all levels.

Consider this excerpt from one company's position statement:

"A viable, disciplined, and organized approach to productivity improvement in all areas and levels of company operations is desirable and necessary towards the accomplishment of Profit, Return on Asset, and Growth goals.

"*Primary* responsibility for productivity improvement rests with responsible top line executives and the leaders of key staff organizations. Unit/Division/Plant managers are responsible for ensuring that an organized and disciplined program supported by goals, objectives, and measures is in effect within their respective units."

The statement, which goes on to describe the duties and authorities of a high-level productivity steering committee and to announce other changes initiated by top management, provides a rationale for a formal productivity improvement effort, clearly fixes responsibility for productivity, and establishes clear expectations regarding the development of productivity programs throughout the organization.

Another company, a manufacturer of automotive products, issued a policy statement that leaves little doubt in the reader's mind of top management's views on productivity and on employees' roles in productivity improvement:

"XYZ Company sells products and services in extremely competitive markets where price, performance, quality and customer service usually determine product preference. To compete in this marketplace environment, the company must be, at a minimum, as productive as its competitors.

"Sustained productivity improvement, therefore, is a basic requisite for achieving the strategic goals of the company. Indeed, productivity improvement is basic even to maintain any presence in many markets. Accordingly, XYZ Company employees have an obligation to shareholders, to other employees and to other constituents to see to it that all company resources are utilized as effectively as possible.

"Our employees are the most important company resource and the resource that can make the largest contribution to improved productivity. How employees utilize their skills, intelligence and energy determines our productivity.

"Because it is so important to all of us that XYZ Company is competitive, it is necessary to establish a formal and permanent productivity improvement process that encourages all employees to apply their skills and ideas to methods of improving productivity. Sustained productivity improvement must become a permanent part of our company culture."[1]

Another excellent example of an effort to communicate management's position on productivity is provided by Beatrice Foods, which issued an eight-page booklet entitled "The Beatrice Productivity Philosophy." Right on the first page, management makes clear its view on the role of people in the effort:

"At Beatrice, our people are the strength of our productivity effort. They help us use our resources more efficiently and effectively. This means working smarter with our tools, our capital, and our human resources."

The booklet goes on to define productivity and to provide a rationale for its improvement:

"Productivity is doing more with the resources we have. It is measured as the simple ratio output/input. . . . Improving productivity within our businesses means finding better ways to do more with the resources we have. With rising energy, labor, and material costs, we must strive to improve all levels of company operations, to maintain our performance in profits and growth."

Finally, the booklet provides advice to the company's managers with respect to building employee awareness through goal-setting, measurement, training, and employee participation.

Good examples of the use of a position statement in service organizations also exist. The Property and Casualty Division of United Services Automobile Association (USAA), a San Antonio-based insurance company, issued in 1986 a document entitled "Productivity Direction Statement."

In an attached cover letter, the division president indicated that the direction statement was ". . . to be used as the guideline for implementing our strategic policy on productivity improvement." Developed by an executive steering committee, this document contains several sections: a statement of the division's mission; a description of its business philosophy; an articulation of its management philosophy; a statement of managerial priorities and key result areas; a list of specific performance objectives; and a series of appendices, including an assessment of past and future productivity efforts, a list of key productivity issues, and objectives for sub-units of the organization.

The section on business philosophy contains this statement:

"Improving productivity within our business means finding better ways to do more with the resources we have. We must strive to improve all levels of company operations and maintain our performance in profits and growth. In short, we need to provide quality products through distribution systems which are customer-convenient and operator-efficient.

"Productivity improvement isn't just working harder, it's working smarter. It means devising a method to get the best return on our investment in people, facilities, equipment and other resources. Our Key Result Areas provide direction for our productivity improvement efforts.

"Productivity growth is a key element in determining improvements in our living standards. It contributes to economic expansion, produces healthy growth in jobs and restrains inflation."

Having discussed the impact and importance of productivity, the Productivity Direction Statement goes on to address, in the management philosophy section, the means by which productivity improvement will be achieved:

"Productivity Improvement will be realized through the employment of contemporary management practices which:

☐ encourage a participative management climate within the P&C (Property & Casualty) Division.

☐ employ advanced technology appropriate to increasing productivity, quality and service excellence.
☐ enhance the physical work environment as appropriate to maximize results.
☐ employ appropriate diagnostic/data-gathering techniques in an ongoing monitoring of our successes and our areas of improvement opportunities.
☐ develop and utilize measures of productivity, quality and service effectiveness and efficiency while not stifling innovation and creativity.
☐ review and challenge the current design of our service delivery systems to ensure the best fit between human resources, technical resources, organizational structures and the service needs of our members, customers, and internal service users.
☐ employ appropriate behavioral science knowledge and skills to ensure the desired level of cooperation, innovation, and support."

This section also amplifies the role of the human resource in productivity improvement: "We at USAA believe people are our most important resource. They help us use all other resources more effectively and efficiently."

COMMITMENT—IS IT VISIBLE?

As we suggested earlier, commitment alone is not enough; that commitment must be made visible to the organization at large through active involvement in the productivity effort. The activities of senior management define the organization's priorities and thus send strong signals to employees. If senior management delegates responsibility for productivity improvement and then backs away from active involvement in the effort, a powerful reinforcing influence will be absent.

An important planning consideration, then, is how to secure the visible involvement of senior management at every organizational entity (company, division, plant, department). That involvement can take many forms—direct communications with middle management and employees, the establishment and publicizing of an organizational productivity objective, the issuance of policy statements that facilitate and support the integration of productivity, the chairing of a steering committee on productivity.

Management's commitment to productivity may be made visible to more than a limited audience. U.S. Gypsum and Foremost McKesson devoted annual reports (1980 and 1983, respectively) to productivity, with their texts focusing on the efforts of the companies' various functional and operating areas in addressing productivity.

One clear indication of management commitment is the amount of resources provided to support the effort. If management will not commit resources to support productivity improvement, how can employees be expected to perceive full management commitment? In an outstanding demonstration of management commitment, Westinghouse Corporation established a seed fund, with continuing contributions for several years in excess of $10 million annually, to fund innovative approaches to productivity improvement that were believed to offer potentially high, but uncertain, returns. Normal financial justifications were not required in order to tap into this fund; a one-page form, describing the nature of the undertaking, was submitted to a management committee for approval. If the committee believed that the initiative held promise and had wider applicability for the company, funds were authorized.

The specific actions taken by senior management will vary and depend upon organizational style and circumstances. What is important is that the explicit and ongoing commitment of top management be made manifest to the organization. Otherwise, the institutionalization of productivity will be an impossible task.

MIDDLE MANAGEMENT COMMITMENT

Developing commitment at the top of the organization is just the start, of course. As the process unfolds, the commitment of middle management and first-line supervision will quickly become an important issue.

Managers and supervisors, like everyone else, tend to resist change. They are comfortable in doing things the old way, and they may not possess the confidence or the skills to change. In addition, they typically labor under a heavy workload and begrudge the time required to support new "programs" that are thrust upon them from above. For these reasons, they are likely to undermine the improvement process—not overtly, but by less-than-enthusiastic participation and support.

Yet these are the very people who must make it happen. The responsibility for implementing much of the process will fall upon their shoulders, and they have direct contact with the workers and significant influence on their behaviors.

Any strategy for implementing a productivity management process will be incomplete if it does not pay considerable attention to middle and first-line management. There are a variety of elements that will increase middle management support:

☐ Building awareness about the need to improve productivity and manage people more effectively
☐ Providing training so that the requisite skills can be obtained
☐ Involving middle and first-line managers in implementation planning, so that they will have ownership of the process

☐ Integrating productivity into the accountability systems for managers
☐ Making top management's expectations explicit

These approaches will be discussed in more detail in later chapters.

EMPLOYMENT SECURITY

If management is committed to productivity improvement *and* is committed to utilizing its human resources to their maximum potential, then the question of employment security cannot be ignored.

To the average worker, unfortunately, productivity is associated with loss of jobs. After all, if the organization is successful in improving productivity, will there not be fewer people required to do the job? In a competitive environment, in the long run, this is normally not true, as will be discussed in Chapter 5. But in the short run, and in the case of shrinking markets or fixed demand, this is indeed a valid concern.

Management is faced here with a dilemma. Productivity improvement is vital to organizational success, but if labor productivity is to be realized, labor costs must be reduced for a given level of output. On the other hand, reductions in workforce as a result of productivity improvement will surely increase employee resistance and ensure the failure of any efforts to win employee commitment to productivity and to involve employees in its improvement.

The company that is truly committed to long-term productivity improvement and sincerely values its human resources ultimately recognizes that it must protect its employees from productivity-related job loss. For a true productivity management process, with its requisite commitment and involvement of employees at all levels, is simply not feasible when employees view productivity improvement as a threat to their jobs.

A number of companies, as a result, have demonstrated their commitment to long-term productivity improvement through a promise to employees that no one will lose his job as a result of productivity improvement. By doing so, they eliminate one of the major impediments to building employee commitment to productivity.

While some companies have made their job security commitment absolute (Lincoln Electric, for instance, guarantees all employees with two-years' seniority at least 30 hours of work per week), most organizations retain the prerogative to reduce the workforce in response to market or economic developments that result in a significant decline in business. There is a price paid, of course, for this provision: management must justify to its employees any layoffs that do occur or risk loss of credibility and trust.

Protection against productivity-related job loss may have short-term cost implications for the firm. To the extent that productivity improvements can only be realized through reductions in headcount, those gains may be deferred temporarily, as excess workers are retrained or as the necessary reductions occur gradually through attrition. Organizations that have made the job security commitment, however, are firmly convinced that the long-term benefits derived from a more involved and committed workforce considerably outweigh any short-term costs.

A number of practices can either ameliorate the cost of a job security commitment or support its implementation:

☐ Manpower planning—the projection of future human resource needs and planning for meeting those needs through productivity improvement rather than hiring.

☐ Retraining and relocation—capitalizing on every opportunity to fill open positions with employees displaced by productivity improvement.

☐ Worksharing—spreading the available work among existing employees through a reduction in working hours.

☐ Buffering—the use of overtime, contracting-out, or temporary employees to meet peak workload requirements and to keep the permanent workforce lean.

☐ Retirement incentives—the offering of special financial incentives to spur early retirement.

Examples of the effective use of these techniques abound. IBM, for example, makes extensive use of manpower planning and buffering to maintain its policy of full employment. Ford Motor Company has invested heavily in retraining programs to support job security commitments negotiated by the United Auto Workers union. Lincoln Electric and Nucor Corporation utilize worksharing to reduce costs while maintaining employment during slack periods. And Ford, DuPont, Exxon, and many others have successfully used retirement incentives to achieve significant reductions in their white-collar staffs.

One researcher identified 30 companies that offer some form of employment security to a substantial portion of their workforces.[2] Some of the better-known of these companies are:

Data General	Hewlett-Packard
Delta Airlines	IBM
Digital Equipment	Lincoln Electric
Eli Lilly	Manufacturers Hanover
Federal Express	Morgan Guarantee
Fort Howard Paper	R. J. Reynolds
Gorman-Rupp	S. C. Johnson
Hallmark Cards	Tandem Computers
Herman Miller	Upjohn

CAN WE GO IT ALONE?

What can a middle manager do when visible commitment and active involvement from the top of the organization are not forthcoming? Many managers can relate to the question asked by the head of a staff department within a fortune 500 company:

"I know that productivity improvement is vitally important to the long-term success of my organization, but it is not our CEO's style to make big pronouncements or to play a visible leadership role in internal company initiatives. Does that mean that I cannot hope to achieve this management process in my department?"

Not all managers interested in productivity are so fortunate as to have their company's leaders blazing the trail. But there is comfort in an indisputable truism: every manager is the top manager in *his* organization. Corporate policies, practices, and culture to the contrary, every manager exerts a significant amount of influence over the activities that take place within his realm. The manager sets priorities, provides leadership, establishes expectations, models desired behaviors, provides feedback, supports (or squelches) employee initiative, sets objectives, and influences employee behaviors through recognition and reinforcement.

The story of the intrepid subsidiary president, whose confrontation with the CEO was related earlier, may yet have a happy ending. He decided that, his CEO's attitude notwithstanding, it was within his power to create a productivity management process within his organization. When last seen, he was huddled with his management team, intently laying out a strategy for change.

Admittedly, the job is easier when the corporate trumpets blow and the CEO produces the tablets of stone. Nonetheless, many managers have succeeded in creating a performance-driven culture within an otherwise unsupportive environment through force of leadership and a commitment to change.

REFERENCES

1. Belcher, John G., Jr., *Productivity Brief 29: Giving Direction to Company Productivity Efforts.* Houston, TX: American Productivity Center, October, 1983.
2. Foulkes, Fred K. and Whitman, Anne, "Marketing Strategies to Maintain Full Employment." *Harvard Business Review,* July–August, 1985.

4

Organizing for Productivity Management

MANAGING THE EFFORT

Once the commitment to the development of a productivity management process has been made, senior management faces a difficult question: how will the effort be managed? In this era of concern about bloated administrative staffs, the answer often is, "We'll manage it through the existing management structure."

But further consideration typically raises more difficult questions. Who will build organizational awareness about productivity? Who will seek to build commitment at all levels of management? Who will explore and analyze the wide variety of improvement techniques available? Who will encourage and facilitate the development of systems to involve employees in productivity improvement? Who will provide guidance and assistance to the various organizational units within the company? Who will serve as a clearing house for information about productivity? Who will serve as management's conscience, reminding senior executives of their role in fostering change?

These tasks could, of course, be parceled out to various managers within the existing structure. The communications department could be charged with building awareness, the human resources department could oversee the employee involvement effort, the controller could be charged with developing productivity measurement systems, and all line and staff executives could be called upon to build management commitment within their respective organizations. And indeed, every part of the organization does have contributions to make to this process.

But who will ensure that all of these activities are carried out in a planned, coordinated fashion? Who will ensure that the awareness-building, measurement, and employee involvement efforts will be integrated, consistent, and mutually reinforcing?

As was suggested earlier, creating a productivity management process is a major undertaking and is an *organizational change* effort. It must be managed if it is to succeed.

THE PRODUCTIVITY MANAGEMENT FUNCTION

In view of the magnitude and complexity of the task, many organizations have created organizational entities to oversee the effort. Management decisions relative to the positioning and manning of this function are among the most important that will be made.

The first decision relates to the organizational *level* of the productivity management function. The importance of this decision cannot be overstated, as it will send a powerful message to the organization. If this responsibility is buried in the bowels of the organization, management's pledge of unswerving commitment will ring hollow. If, on the other hand, the newly-created position occupies a senior place on the organization chart, management's commitment to productivity will be undeniable. The title Vice President, Productivity speaks volumes about top management's position on the subject.

Many companies in a wide variety of industries have recognized the importance of the productivity management function by according it the stature of a vice presidency. Among these are Armco Corporation, Bank of America, Sentry Insurance, Harris Corporation, American Hospital Supply, and Westinghouse. Numerous others have created corporate positions carrying such a title as Director of Productivity.

The suggestion of creating a corporate productivity management function often evokes a response such as, "That's all we need, another corporate bureaucracy." That very concern has led to the prevalence of the "one-man band," the productivity executive with no direct staff. This individual's primary role is to serve as the company's "champion" for productivity, advising top management and encouraging productivity efforts throughout the organization. As will be discussed later, this individual is often supported by a network of part-time productivity coordinators located in the various operating units of the company.

Where large productivity staffs do exist, they typically result from a consolidation of existing corporate resources under the productivity banner. This approach is typified by Westinghouse, which brought together existing quality, industrial engineering, and technical staffs into a corporate productivity center with 140 employees. About half of the center's funding is provided by the corporation, to be spent on activities having a long-range payoff, with the remainder provided by the internal users of the center's services.

Another decision surrounding the creation of a productivity management position relates to the *functional location* of the position. Should it be housed in Engineering? Human Resources? Finance? While the productivity management function is commonly found in all of the aforementioned areas, none of them represents an ideal home.

The association of the productivity effort with a given functional area will give the effort an inevitable (and perhaps unintended) caste. An engineering-based productivity effort will be perceived by the organization as a purely technical endeavor and will raise fears of technological displacement and work measurement. If housed in the human resources function, the effort will likely be viewed as a "soft" motivational exercise. And a finance-driven effort will likely be viewed as yet another cost-reduction and head-cutting exercise.

The ideal home for the productivity executive is in his own box, reporting directly to the chief operating officer (or the head of whatever unit he serves). This positioning serves three important purposes: It avoids the functional taint, ensures that the position will carry the weight of authority, and provides an excellent demonstration of top management commitment to productivity.

Companies have been known to make some unusual decisions with regard to the home of the corporate productivity executive. This species has been found in such strange habitats as the public relations department and the office of the general counsel. It is interesting to speculate upon the nature of the signal sent to the employees of those organizations.

The final question surrounding the establishment of a productivity management position is that of *responsibilities*. Following is a list of responsibilities for which productivity executives are often held accountable:

□ **Advising senior management regarding their role in the effort.** Given the importance of visible top management commitment, as discussed in Chapter 3, it is important that the productivity manager promote their involvement and develop means to make their commitment visible to the organization.

□ **Building organizational awareness of productivity.** The productivity management function must bear primary responsibility for educating the organization at large about the nature of and rationale for productivity improvement. Some techniques for building awareness will be discussed in Chapter 5.

□ **Building commitment at all levels to productivity.** Enlisting, through education and persuasion, the support of key managers and informal leaders at all levels serves to accelerate the pace of change.

□ **Disseminating guidelines or expectations to the various organizational components.** The productivity manager should communicate, clarify, and support management's expectations regarding the broad direction of improvement efforts. Involving employees in these efforts, for example, might be an important expectation of top management.

□ **Monitoring progress and identifying problem areas.** The productivity manager should regularly evaluate progress in the development of the management process and identify those areas where additional management attention is required.

☐ **Serving as a resource to the organization.** Supporting the effort through the provision of guidance, advice, and assistance is a vital role for any productivity management function.

☐ **Serving as a gathering point for information about productivity improvement techniques.** Many techniques have wide applicability throughout the organization, and the productivity management function can play an important role as a source of information about these techniques.

☐ **Transferring successes from one part of the organization to another.** In a large organization, change will be impeded if there does not exist an effective mechanism for identifying successful initiatives and transferring the learnings to other organizational units.

☐ **Serving as an organization's focal point and champion for productivity.** In general, the productivity manager should serve in a leadership role by maintaining high visibility and promoting change.

Some examples of job descriptions for productivity managers can be found in Appendix A.

THE PRODUCTIVITY EXECUTIVE

Once the specifics of the productivity management function have been established, concern naturally turns to the qualifications of the person who will manage the function. This issue should be carefully considered, as the success or failure of this individual may have major ramifications for the future success of the organization.

Ideally, the productivity executive should possess the following characteristics:

☐ **An understanding of the business.** Since the productivity executive must interface with line management throughout the company and will contribute to the identification of productivity improvement opportunities, an understanding of the dynamics and technical aspects of the business is an important asset. For this reason, many companies (Armco Corporation and American Hospital Supply are examples) fill this position with a general manager of one of the operating units.

☐ **The respect of the organization.** Since the productivity executive typically has no direct authority over other managers, he must build commitment and rely on persuasion to achieve his objectives. This is most assuredly not the appropriate position for the ineffective, abrasive, or contentious manager. Amazingly, some companies have used this position as a final resting place for the near-retirement executive who has been moved out of his previous position to make way for an energetic up-and-comer. It is awful to contemplate the signal sent to the organization by this maneuver.

☐ **Quantitative skills.** Since measurement (Chapter 6) is a key element of the productivity management process, the productivity executive that is not comfortable with numbers will lack an important skill.

☐ **Communications skills.** The productivity executive will spend a larger proportion of his time communicating with others than will most other managers. He must educate, inform, persuade, and cajole. An individual who is not a good communicator will suffer a serious handicap in his efforts to bring about change.

☐ **A human resource orientation.** Perhaps the most important qualification of all is an abiding belief in the primacy of the human resource. People are the key to productivity improvement, and the creation of an environment in which people can and will contribute to the maximum of their capabilities is a prerequisite for success. The productivity executive must therefore champion efforts to develop and involve employees at all levels and in all parts of the organization. The autocratic manager who attributes his past success to his ability to whip people into line through the exercise of authority is not the right person for the job.

If the person possessing these qualifications sounds like the All-American Executive, it is because he is. How can an individual be expected to manage a major organizational change effort if he is not technically competent, multi-skilled, people-oriented, and influential? Companies with savvy and foresight tend to select their high-caliber, fast-track people to staff the productivity management function.

THE PRODUCTIVITY STEERING COMMITTEE

Another organizational structure commonly found in companies committed to productivity is the Productivity Steering Committee. While this structure, consisting of senior executives from various functional areas, serves for some organizations as an alternative to a productivity executive, most companies view the steering committee as a complement to the productivity manager.

The functions of the productivity steering committee are to provide direction to the productivity executive, if one exists, and to provide overall governance of the effort. The committee, which typically meets monthly or quarterly, establishes policy, sets goals for the effort, reviews progress, and provides the authority to modify organizational systems to support and reinforce productivity improvement.

In addition to providing the clout to ensure that things happen in the organization, the senior management steering committee provides an excellent demonstration of top management's commitment to and involvement in the productivity management effort.

THE PRODUCTIVITY MANAGEMENT NETWORK

In a large, multi-location company or division, the task of the productivity executive is indeed a daunting one. For how can one person effectively influence the course of productivity in dozens, or even hundreds, of organizational units?

The productivity management effort in these organizations benefits greatly from a cadre of productivity coordinators strategically located throughout the company.

In the typical application of this approach, these coordinators do not report directly to the productivity executive, but are attached to the operating or staff unit in which they reside. They do, however, serve as an extension of the productivity executive, advocating, supporting, and encouraging the productivity management philosophy being promulgated by the firm. Such an infrastructure greatly facilitates communications and dissemination of corporate guidance and ensures that the improvement effort is receiving attention throughout the organization.

The network approach serves another important purpose: It places responsibility for developing the productivity management process in the hands of the operating units, thus avoiding the unfortunate perception that productivity management is a staff function.

An early proponent of the productivity network organization was TRW, which began appointing productivity coordinators in 1980. Coordinators were sought at the group, division, and plant levels, and in corporate administrative functions as well. The positions were part-time (with some exceptions) and were staffed with high-potential employees. The ultimate objective was to have 1,200 productivity coordinators in place worldwide.

To ensure that its productivity coordinators were properly trained in the principles of productivity management, TRW established a "Productivity College." The agenda for this three-day training program, which was attended by all newly-appointed productivity coordinators, is presented in Figure 4-1.[1]

With an increasing awareness of the magnitude of the task, the network organization has grown in popularity in recent years and can now be found in such companies as Boise Cascade, Wisconsin Gas Company, and Warner-Lambert.

MULTIPLE STRUCTURES

An excellent example of an organization that has created multiple structures to meet its particular needs is Honeywell's Aerospace and Defense Group.[2] With 20,000 employees spread across the country in several divisions, the A&D Group recognized the need for dedicated organizational en-

(text continued on page 40)

Day One

Productivity and TRW: A Conceptual Setting
Productivity—What is it? What are the benefits?

A process model for change:
- Awareness
- Planning and analysis
- Assessment and preparation
- Training and implementation
- Evaluation and maintenance

Exercise: Best/Worst characteristics of an organization. What makes
organizations functional and dysfunctional?

Overview of employee participation and team building.

Quality Circles

Quality of worklife/employee involvement techniques.
- Communications programs
- Survey feedback
- Suggestion systems

Day Two

Implementing a team management system.

Introduction to measurement.

Measuring the knowledge worker.

Exercise: Measurement at TRW.

Brainstorming exercise: what can be done to improve productivity with
little or no additional resources?

Self-evaluation exercise of the productivity climate at each participant's
operation.

Day Three

Case Study: Richmond Automotive Parts facility

Productivity—a strategic approach

Case Study and discussion groups—Lincoln Electric

Wrap-up and summary

Figure 4-1. TRW "Productivity College" agenda.

tities to communicate, foster, and support management's productivity and quality philosophies.

This recognition led to the creation of a number of structures with different purposes:

☐ A Productivity Advisory Board consisting of senior Group management and several outside experts provides overall guidance to the effort.
☐ A Productivity/Quality Center, with approximately 20 employees, provides internal consulting services in such areas as office automation, work analysis, CAD/CAM, and just-in-time production.
☐ A Management Development Center develops and implements training programs to improve the quality of management.
☐ A Human Resource Center supports employee involvement activities through such services as facilitation training and quality circle implementation.
☐ Ten Professional Councils—one for each major functional group (Engineering, Purchasing, Finance, etc.)—bring together functional directors from across the divisions to exchange information, discuss common problems, and devise functional productivity strategies.

The multiple structures created by the A&D Group obviously require a significant commitment of resources, manpower, and time. But an organization better positioned to foster change would be hard to find.

A CAVEAT

There is a potential trap associated with the creation of a productivity executive, and it must be avoided; it is imperative that the productivity executive or productivity coordinator not be perceived as having responsibility for productivity *improvement*. Such an outcome would be inconsistent with an important element of the productivity management philosophy: that every member of the organization is responsible for productivity improvement as part of their day-to-day jobs. Ownership of and commitment to that responsibility is vital to the success of the productivity management initiative.

Rather than having responsibility for productivity improvement, the productivity manager is responsible for overseeing the *development of the productivity management process*.

REFERENCES

1. Ruch, William A. and Werther, William B., Jr., "Productivity Strategies at TRW." *National Productivity Review*, Spring, 1983.
2. *Case Study 34: Honeywell Aerospace and Defense Group.* Houston, TX: American Productivity Center, 1984.

III

INTEGRATING AND REINFORCING PRODUCTIVITY

5

Awareness and Accountabilities—Prerequisites for Change

THE TECHNIQUES TRAP

Once the commitment is present and organizational decisions have been made, it is tempting to begin experimenting with various techniques designed to bring about productivity improvement. Quality circles, work simplification, value analysis, office automation, and a host of others may be tried in a burst of organizational enthusiasm and zeal.

Not that these techniques should be avoided; on the contrary, improvement techniques are a critical part of the improvement effort. The problem arises when techniques are instituted before the organization is prepared, before the groundwork has been laid to provide a context within which the techniques have meaning. If productivity has not been established as the encompassing framework for the various techniques, then each technique will be perceived by the organization as a program, an isolated effort that is not an integral element in an organized, cohesive strategy. An organization whose productivity improvement efforts consist solely of techniques cannot expect to achieve the productivity culture; rather, it can expect productivity to be addressed by the workforce only within the context of the organization's present repertoire of techniques. It will lose the power of a culture in which productivity improvement is pursued on a day-to-day basis as an operating norm of the organization.

It is critical, therefore, that management not become enamored of techniques to the extent that these techniques become ends in themselves. Rather, they should be viewed as tactics which, at the appropriate place and time, support the overall strategy of integrated, continuing productivity improvement.

But how does management prepare the organization so that improvement techniques can be most effective and productivity can be successfully integrated into other management systems and practices? The process should start with organizational awareness-building.

BUILDING AWARENESS

What does productivity mean to employees? Are they fully cognizant of its implications? Can their full cooperation in the productivity improvement effort be anticipated?

An organization that tries to realize significant productivity improvement without the participation and support of its employees is working against itself. It doesn't make much sense to embark upon a major undertaking when the bulk of the organization misunderstands—or worse yet, resists—the object of that undertaking.

To many employees, the word "productivity" carries a negative connotation. Too often it conjures up images of work speed-ups and headcount reductions. The connection between productivity and the health and prosperity of the organization and its employees is generally not fully appreciated.

One of the management's first tasks, then, is to raise the level of awareness in the organization at large (assuming management's own awareness needs have been attended to, as suggested in the preceding chapter). The rationale for productivity improvement needs to be made clear and salient to every employee.

An effective way to begin the awareness-building process is through dissemination of the productivity philosophy statement described in the previous chapter. This statement would explicitly recognize the implications of productivity to the organization and to the individual. It would establish the priority for productivity and define responsibilities. It would also recognize the employees' role in productivity and would legitimize the organization's human resource strategy. By itself, this statement will not cause significant change to occur, but it can serve as a tangible document to launch the effort.

The policy statement, however, should only be one element of a more intensive awareness-building process. As long as misconceptions about productivity are widely held by the workforce, any effort to build a productivity-driven culture must ultimately fail. Among the more deadly (and all-too-common) misconceptions are these:

☐ **Productivity is a management code word for "working harder."** While some gains could undoubtedly be achieved in most organizations through more disciplined application of effort by employees, the real leverage for improvement lies in collaborative problem-solving, better ways of organizing and carrying out work, and full utilization of the skills and capabilities inherent in the workforce. The often heard phrase, "Productivity means working smarter, not harder," says it well!

☐ **Productivity improvement means loss of jobs.** While the concern may have some validity in the short run (and therefore needs to be managed, as suggested in the previous chapter), the long run implications are actually to the contrary: productivity improvement *creates* jobs. The company that is most successful in improving productivity, relative to its competitors, will gain market share and will thus expand and grow. In addition, if the rate of a company's productivity improvement exceeds the increases in the costs of its inputs, the prices of its products or services can actually fall, and overall levels of demand will increase. In actuality, it is the *lack* of productivity improvement that causes loss of jobs.

☐ **Capital investment represents the primary opportunity for productivity improvement.** While significant improvement opportunities through capital investment certainly exist, the major opportunity for many organizations lies in more effective utilization of the capabilities of its human resources. Most companies rigorously pursue capital investment opportunities on an ongoing basis, and many have largely capitalized on the potential that exists in that arena. Few companies, however, have capitalized on the improvement potential inherent in their people, and the opportunity here is enormous. Productivity improvements of 50% and more through human resource oriented efforts are not uncommon, and instances in which a company, through its human resource practices, has achieved a productivity level twice that of its competitors have been documented.

☐ **Productivity applies only to the factory.** Historically, the focus of productivity programs has been on the blue-collar worker. The factory work force has been squeezed to such a degree that even in many manufacturing organizations they are now in the minority. And the factory does not even exist in many sectors of our service-oriented economy. The productivity of white-collar workers—particularly professional, technical, and managerial employees—represents a major opportunity for most organizations.

☐ **Productivity benefits only management.** The fact is that long-term productivity improvement benefits *all* constituency groups. Yes, management benefits from business growth and improved returns. But employees benefit from more secure jobs and an improved capability of the organization to increase compensation. Consumers benefit from lower prices and a higher standard of living. The nation, and therefore its citizens, benefits from a stronger economy and a better competitive position in international markets.

VEHICLES FOR AWARENESS-BUILDING

As in organizing for productivity, there is no single correct method of building awareness. A variety of approaches—written communications, professional presentations by outside experts, senior management communications to employees through video tapes or face-to-face meetings, supervisor-employee discussion sessions—can be effective, depending upon the circumstances.

A number of companies have invested considerable resources in the development of video tapes as a means of reaching a large number of employees. Warner-Lambert Company, for example, showed a professionally-produced video tape to all of its 40,000 employees in small groups. The tape featured a well-known media personality, who provided a macroeconomic context for productivity; the company's CEO and president, who described the implications of productivity to the company's future well-being; and several vignettes of Warner-Lambert employees enthusiastically recounting their contributions to company productivity. Other companies utilizing video tapes for broad-based awareness-building include Anheuser-Busch, Lear-Siegler, and Northrop Corporation.

Booklets and pamphlets, such as "The Beatrice Productivity Philosophy" referenced in the preceding chapter, can also be effective vehicles for mass communications. Another example of this approach is provided by National Semiconductor Corporation, which issued a pamphlet entitled QUEST (an acronym for Quality Enhancement Strategy). Commencing with a discussion of the importance of productivity and quality to the company's future success, the brochure goes on to describe the nature, objectives, and components of the improvement program, emphasizing the critical role of employees in the effort.

One of the slickest examples of an awareness brochure is provided by the Lockheed Missiles and Space Company (LMSC), a subsidiary of the Lockheed Corporation. Entitled "Productivity—A Total Commitment at LMSC," the booklet contains 28 pages and is as slick as any marketing brochure or shareholder annual report. Starting with brief messages from the company's chairman, president, and director of productivity, the brochure is divided into several sections, each with its own message:

☐ The Total Commitment. Describes the history of the company's productivity effort.

☐ Tapping the Talent Pool. Recognizes the contributions of its people to the success of the company's endeavors.

☐ Ideas—Positive and Timely. Reinforces the importance of employee ideas and sanctions employee input to a wide variety of opportunity areas.

☐ Strategy—Everyone's Involved. Emphasizes the strategic planning underlying the productivity effort and affirms management's commitment to participative management.

☐ Exploiting Our Full Potential. Describes the support provided to its employees in making contributions.

☐ It Happens Every Day. Describes a variety of techniques being utilized to improve productivity.

☐ Setting the Example. Promotes cost reduction as an integral element of the productivity improvement process.

☐ Measuring Success. Highlights the savings achieved to date through the improvement effort.

☐ Looking to the Future. Recognizes the potential for enormous gains and reaffirms management's commitment to utilize the talents of its people to the maximum extent possible.

Awareness-building vehicles need not be limited to traditional communications devices. Foremost-McKesson piqued employee interest in the subject by conducting a productivity poster contest.[1] The objectives of the contest, according to the company, were twofold: to encourage creative thinking about productivity by employees and to reinforce the notion that productivity is not just a management concern. The winning poster was distributed throughout the company for continuing service in raising awareness.

Given the magnitude of the task in a large organization, a multifaceted approach to awareness-building is generally desirable. One organization that effectively addressed this need is the Operations Division of Manufacturers Hanover Trust, a 7,500-employee operation.[2] An awareness subcommittee (of the divisional productivity committee) developed and employed a variety of vehicles:

☐ Presentations to all management employees of the results of an organization-wide survey of employees' attitudes toward productivity

☐ Inclusion in the quarterly house magazine of stories about employees who had made significant contributions to the productivity effort

☐ A new publication dedicated to productivity success stories

☐ Productivity posters located throughout the division's offices

☐ The printing of a slogan, "You are the key to productivity," on a variety of communications documents

☐ High-visibility awards, such as trophies and certificates, to individuals and groups that had made significant productivity contributions

AN ON-GOING TASK

Continuous attention to organizational awareness is a prerequisite for an effective productivity management process. Awareness-building is not a task that is completed and then set aside. Constant communication about productivity—and the organization's progress in improving it—is a necessity if an organizational focus on productivity is to be maintained.

As a vehicle for this on-going awareness and communications process, a number of companies publish productivity newsletters for regular distribution to the workforce. One of the better examples is "The Productivity Review," published by the Federal Systems Group of Sanders Associates. This monthly publication carries articles on the productivity challenges faced by the organization, formal improvement efforts undertaken, ideas from employees, and tips for overcoming common barriers to productivity improvement.

Another good example of an ongoing awareness vehicle is a newsletter entitled "Productivity," published by Northern Telecom, a Toronto-based company with extensive operations in the United States. Published by the company's Corporate Productivity Center, some recent articles included the following:

- ☐ "What is Productivity?" A detailed discussion of productivity and its implications.
- ☐ "Just in Time." A description of the startling improvements (assembly time for one product reduced from seven and a half days to 88 minutes) achieved through the application of just-in-time principles.
- ☐ "Supervisors Key to Worker Productivity." An article by a Harvard Business School professor advising that supervisors be actively involved in planning an employee involvement effort in order to reduce their resistance.
- ☐ "Innovation + Teamwork = Productivity." A case study describing the successful development of an advanced new product through a coordinated team effort.
- ☐ "Technological Advances Maintain Employment." Excerpts from a speech by the company's chairman on the critical importance of technological advances to a modern, evolving society.
- ☐ "Measurement is Good Management." A discussion of productivity measurement and its value in promoting productivity improvement.
- ☐ "Nominal Group Technique." Information on a brainstorming technique that is useful in generating ideas and achieving group consensus (this technique is discussed in Chapter 8).

MAKING RESPONSIBILITIES EXPLICIT

In addition to building awareness, the organization seeking to develop a productivity management process must also act early to develop explicit responsibilities and clear accountabilities for productivity improvement among its management personnel.

These actions may, at first, seem to be unnecessary. Just ask any manager or supervisor—he will affirm his responsibility for productivity. After all, what manager would have the temerity to suggest that he is not responsible for improving productivity in his organization?

Unfortunately, however, this expressed acceptance of responsibility too often is little more than lip service. The typical manager or supervisor spends very little time in assessing productivity improvement opportunities, planning and implementing more effective productivity management practices, and developing his subordinates to more fully contribute to the performance of his organization.

This behavior, of course, is not surprising; the typical manager is simply not held accountable for productivity improvement. He is held accountable for fighting fires, meeting his budget, and following his superior's orders.

The issue of responsibilities for productivity improvement, therefore, cannot be ignored. For a true productivity management process cannot be achieved when accountabilities are weak and responsibilities are implied and nonoperative. Productivity cannot be an operating norm until all managers accept—and act upon—the responsibility for productivity as an integral element of their day-to-day jobs.

A variety of actions help establish productivity as an explicit management responsibility, such as:

☐ Defining these responsibilities in the productivity policy statement.
☐ Setting improvement goals and holding managers accountable for their achievement. TRW, for example, is well-known for its process of "cascading" goals, including those related to productivity, down through various levels of the organization.
☐ Inserting specific productivity language in job descriptions.
☐ Establishing productivity performance as an explicit element of the performance appraisal process. The typical performance review form does not even contain the word "productivity."
☐ Clearly establishing productivity performance as a criterion for promotion. Top management must send the message, "If you want to get ahead here, you must demonstrate the ability to improve productivity."

Ultimately, the responsibility for productivity must be made explicit for all employees, not just managers. Acceptance of this responsibility by nonmanagement employees comes gradually, as the awareness-building and employee involvement (Chapter 7) processes pervade the organization.

PREREQUISITES FOR CHANGE

Initiatives to make responsibility explicit and to increase organizational awareness often get short shrift in company productivity programs. They do not provide the immediate and measurable returns offered by investments in new technology, improvements in cycle times, and reductions in work force. But without them, the organization is doomed to a never-ending cycle of one-shot productivity improvement, initiated by senior management or staff specialists, and resisted by the workforce.

When the organization is aware and supportive, and responsibilities are explicit and unambiguous, powerful reinforcing and sustaining mechanisms are in place. Without them, productivity can never become institutionalized, and a productivity management process can never exist.

REFERENCES

1. "Foremost-McKesson Productivity Poster Contest," *Productivity Letter*, Houston, TX: American Productivity Center, July, 1983.
2. Lambert, Richard J., "Productivity Awareness at Manufacturers Hanover Trust." *National Productivity Review*, Summer, 1983.

6

Productivity Measurement

MEASUREMENT AS A REINFORCING MECHANISM

Measurement is integral to the productivity management process. If productivity is to be integrated into the organizational culture, a vehicle for monitoring progress, providing feedback, setting quantifiable objectives, and evaluating managerial performance is a *sine qua non*.

The status of productivity as a strategic issue to the organization suggests that its monitoring is essential. How else can an organization, or one of its components, determine whether it is a Company A (improving profitability through productivity) or a Company B (improving profitability through price recovery)? Productivity measurement also aids the organization in explicitly relating productivity to its other strategic objectives; productivity improvement may be the primary means of achieving growth in market share, for example. The existence of a reliable productivity measurement system enables the organization to sharpen its strategic plans through the establishment of targeted productivity improvement levels related to the achievement of a specific strategic objective.

Apart from its strategic usefulness, productivity measurement performs a number of valuable reinforcing functions. Some of the useful functions of measurement include the following:

☐ **Building awareness.** Highly visible and often-referenced measurement systems help to maintain organizational focus and communicate management's interest and concern about productivity. An engineering organization, as will be described later in this chapter, greatly heightened employee interest in productivity by graphically displaying its measures in a conference room specially built for that purpose.

☐ **Assessing problem/opportunity areas.** Productivity measures facilitate identification of areas where management attention is needed. A condition of flat or declining productivity can only be ascertained through measures.

☐ **Providing a mechanism for feedback.** Without feedback, an organization cannot learn and improve. By feeding back measurement data, employees can enjoy a sense of accomplishment, can learn from success, and
can be motivated to overcome periods of inadequate performance.

☐ **Facilitating integration.** Measurement facilitates the process of integrating productivity into other organizational systems; quantifiable goals can
be set, productivity improvement can be budgeted, and reinforcement
through the reward system can be accomplished with greater objectivity.

Once productivity measures are developed, their integration into the financial reporting system should be accomplished. Productivity measures are
not something that should be examined by management as a side issue to
financial performance, but should rather be viewed as an integral *determinant* of financial performance. The impact of productivity on the bottom
line should, therefore, be a routine element in any analytical presentation of
financial results. Otherwise, productivity will never escape its "program"
taint.

Productivity measurement should also be integrated into the budgeting
and planning systems of the organization. Projections of productivity measures, along with their impact on costs and profits, belong in every budget
and long-range financial plan. If they are not there, productivity has not
been institutionalized.

In spite of the fact that measurement plays a vital role in managing productivity, measurement systems in most companies are weak. The traditional financial measures of performance—profits, return on assets, etc.—
are not valid indicators of productivity because they are also affected by
price recovery (Chapter 1). In addition, the productivity measures that do
exist in organizations are typically limited to the labor input and are thus
inadequate.

While the development of a measurement system should be a priority in
the productivity management process, the *improvement* of productivity
should not be deferred pending the availability of satisfactory measures.
The view that it is pointless to try to improve productivity before it can be
measured reflects a failure to appreciate the full implications of productivity
to the organization. To date, no physical law has been discovered that renders productivity improvement impossible without measures.

A COMPLETE PRODUCTIVITY MEASUREMENT SYSTEM

Given the valuable reinforcing capabilities of measurement, organizations with effective productivity management processes generally make
rather comprehensive use of measurement. Measurement is used at the

working level by employees and supervisors to monitor and regulate performance on a day-to-day basis. Profit-center and general managers utilize measurement to interpret financial results and assess the health of their business. Administrative and support groups use measures that are designed to evaluate their contribution to the organization's broader objectives. Senior executives use measurement to improve the effectiveness of strategic planning and decision-making. National and industry productivity data are examined as comparative benchmarks.

It is important to recognize that different approaches to productivity measurement are necessary to serve different purposes. The types of productivity measures used on the factory floor or in a clerical operation are of little value (or may even be detrimental) in a knowledge-worker environment. The measurement system developed by line management for day-to-day control purposes is probably inadequate for strategic decision-making. And measurement systems used for strategic business planning will likely be incomprehensible to the hourly worker.

Above all, measurement systems should be designed to be useful to the organizational unit being measured. Perhaps self-evident, this principle is nonetheless often violated. Far too many corporate staffs have been observed sending letters to the operating units saying, in effect, "Thou shalt measure productivity thusly and report the numbers to us by the fifth working day of each month." More often than not, the measures dictated are viewed by operating management to be of no use to them in managing their business; their calculation therefore becomes just another bureaucratic chore.

PARTIAL VS. TOTAL PRODUCTIVITY

Measures can be characterized as either partial or total productivity measures. Partial productivity measures are derived by dividing the total output of the organization by a single input:

$$\frac{\text{Output}}{\text{Labor}} \qquad \frac{\text{Output}}{\text{Materials}}$$

$$\frac{\text{Output}}{\text{Capital}} \qquad \frac{\text{Output}}{\text{Energy}}$$

Partial productivity measures are useful but have a shortcoming: one partial measure can be improved at the expense of another (the substitution effect). A common example is the installation of labor-saving capital equipment; labor productivity should improve as a result, but capital productivity may well decline.

The ultimate indicator of an organization's effectiveness in addressing productivity is a total productivity measure. Total productivity is defined as total output divided by the sum of all the inputs:

$$\text{Total Productivity} = \frac{\text{Output}}{\text{Labor} + \text{Materials} + \text{Capital} + \text{Energy}}$$

As suggested earlier, no single productivity measure, or set of measures, is appropriate throughout a large and complex organization. The specific approach to measurement should be dictated by the organizational level and the use to which the measures are to be put.

MEASUREMENT IN A REPETITIVE WORK ENVIRONMENT

Productivity measures at the operating level—the plant or department—are usually partial measures and are used for basic control purposes. That is, they are used to monitor performance on a relatively frequent basis (daily, weekly, or monthly) so that operating problems can be identified and dealt with in a timely fashion. They also have value in tracking longer term trends; annual goals, for example, are often set around these measures. Most organizations have productivity measures of this type, but their quality and breadth of coverage are frequently inadequate.

Since productivity in the factory is basically a physical concept, relating quantities of output to quantities of inputs, partial productivity measures at the operating level in a manufacturing or clerical environment ideally should be constructed with physical variables, i.e., units of output compared to units of input. Some common examples of physical partial productivity measures are provided in Table 6-1.

It should be noted that physical measures are susceptible to distortion from changes in product (or service) mix. The typical manufacturing plant,

Table 6-1
Physical Productivity Measures

Labor	$\dfrac{\text{Units Produced}}{\text{Man-Hour}}$	$\dfrac{\text{Transactions Processed}}{\text{Employee}}$
Materials	$\dfrac{\text{Tons Out}}{\text{Tons In}}$	$\dfrac{\text{Units Shipped}}{\text{Units Scrapped}}$
Capital	$\dfrac{\text{Units Produced}}{\text{Machine Hour}}$	$\dfrac{\text{Tons Shipped}}{\text{Trucks Employed}}$
Energy	$\dfrac{\text{Tons Processed}}{\text{BTU's Used}}$	$\dfrac{\text{Items Completed}}{\text{Kilowatt Hour}}$

for example, makes several different products, and those products often require differing amounts of inputs to produce. If a plant measures its labor productivity in terms of units of output per man-hour, a shift in production mix towards units that inherently require more man-hours to produce will cause the productivity measure to decline (fewer units will be produced with the given man-hour input). This provides a false signal, as the organization has not used its resources less effectively in producing its outputs.

This phenomenon can be overcome through a weighting mechanism. Rather than simply adding up disparate units of output, each distinct type (or family) of output is weighted (multiplied) by a factor which equates it with all other types of output in terms of the input being measured. In a simple example, a plant that produced two products, one of which requires twice the labor input of the other, would weight the former by a factor of two, the latter by a factor of one. Obviously, different weights may be required when measuring the productivity of materials or some other input. In organizations utilizing a standard cost system, a ready-made weighting system may be available in the physical standards.

A standards-based weighting system used at one time by Emerson Electric is instructive. At each of Emerson's plants, a weighting factor was calculated for each product (or product line) by dividing that product's direct labor standard by a constant factor which represents the relationship between the plant's total output and the total direct labor hours worked during a base year. Sample calculations are illustrated here for development of weighting factors:

Base Year

Units Produced	10,000
Direct Labor Hours	20,000
Average Hours per Unit (Constant Factor)	2.0

Weighting Factors

	Direct Labor Standard	Constant Factor	Weighting Factor
Product A	4.0	2.0	2.0
Product B	3.0	2.0	1.5
Product C	2.0	2.0	1.0
Product D	1.0	2.0	0.5

The plant then measures its labor productivity in terms of weighted units produced per direct labor hour worked. The output side of the equation is simply the aggregation of the units of each product produced multiplied by the appropriate weighting factor. Mix changes are automatically compensated for and do not distort the measure.

It should be noted that the weighting factors must remain fixed, even if the labor standards change over time. If the weighting factors are recalculated each time the standards change, the ability to track the productivity trend will be lost. The base against which each year's productivity performance is calculated will be different, and the long-term trend will not be discernible.

Where physical measures are not feasible or practical, financial (dollar-denominated) variables may serve as surrogates for the physical inputs or outputs. Financial measures are particularly useful when measuring capital productivity (since buildings, land, and equipment cannot be aggregated in a physical sense) and materials productivity (since different forms of materials also may be impossible to aggregate physically).

When a productivity measure contains both a financial and a physical component, e.g., sales/employee, it is imperative that the financial component be deflated, or stated in constant-dollar terms. To fail to do so renders the measure useless as an indicator of productivity. In the example cited, sales will inevitably increase over time due to the impact of inflation. But inflation-related increases in output do not represent productivity improvement; they have nothing to do with how effectively we have utilized our resources in producing our outputs.

Pure financial measures, such as sales/payroll, may be viewed as self-deflating, if one is comfortable with the assumption that both the numerator and the denominator are inflating at equal rates. Since the real world is rarely so cooperative, it is generally wise to deflate both variables separately to reflect their differing rates of change.

The ideal deflator is an internally-developed price or cost index that reflects the company's specific experience. A price index would be developed by dividing each period's average selling price by the average selling price in the base period. The productivity measure for each period is then divided by that period's index in order to obtain the deflated result. A simple example is presented in Table 6-2. Lacking appropriate internal data, the organization may resort to one of numerous indexes of a more general nature that are published by the federal government.

Table 6-2
Deflation Through Indexing

Year	Actual Sales	Price Index	Deflated Sales
1	$1,000	1.00	$1,000
2	$1,050	1.08	$ 972
3	$1,100	1.14	$ 965
4	$1,250	1.20	$1,042
5	$1,400	1.30	$1,077

The inflation problem becomes particularly acute when measuring capital productivity. A company's fixed assets are valued by the accountants at the original purchase price, adjusted for depreciation. Since these assets have typically been acquired over many years at significantly different price levels and have likely been depreciated through some accelerated method in order to reduce taxes, their book value is almost certainly not valid as an input measure. As an example, two identical pieces of equipment may carry vastly different book values because they were acquired years apart. Financial measures of capital productivity therefore generally require that the value of the capital input be restated at a constant price level (e.g., 1986 prices) in order to accurately reflect the true physical input contributions of the organization's capital stock.

A comprehensive productivity measurement system at the operating level will consist of partial measures for all the major inputs under the control of that operating unit. A simple measure of total productivity may also be appropriate, and can be derived by expressing the partial measures in index form (base period equals 1.0) and aggregating them in such a fashion that each partial's weight in the total measure reflects that input's proportion of the total cost structure. In other words, if labor represents 20 percent of total cost, the labor partial measure should account for 20 percent of the total productivity measure. An example is provided in Table 6-3.

Table 6-3
Total Productivity Derived from Partial Measures

	Partial Index	Weight	Total Index
Labor	1.06	.20	.21
Materials	0.98	.40	.39
Capital	1.14	.30	.34
Energy	1.02	.10	.10
Total			1.04

MEASUREMENT FOR FINANCIAL AND STRATEGIC PLANNING

If productivity is to be an integral part of strategic and financial planning, a means of evaluating and projecting total organizational productivity performance is requisite. The operational measures discussed in the previous section are of limited usefulness in strategic decision-making and long-range planning; they simply provide too much detail.

The problem, then, is to bridge the gap from physical measures of operational control to the "big picture" needs of higher level management. Given the strategic importance of productivity, senior management must have a vehicle that will clearly indicate how productively the organization is utilizing all of its resources. Management must also have a means of analyzing the income statement in order to determine whether profit growth is being fueled by productivity or price recovery. As indicated earlier, continuous over-recovery of input cost increases may lead to serious competitive problems. Finally, productivity measures are needed for effective strategic planning; a strategic business plan is incomplete if productivity improvement is not an explicit and integral element of the plan.

One approach is illustrated by a report from a total performance measurement system developed by the American Productivity Center (see Table 6-4). The system is in use in several companies, including Ethyl Corporation, Olin Chemicals, and Sentry Insurance. This analysis may be produced at the corporate level, or reports may be issued for each plant or division and aggregated to the corporate level.

Table 6-4
Performance Report

Input	Performance Index			Effects on Profits		
	Profit-ability	Produc-tivity	Price Recovery	Profit-ability	Produc-tivity	Price Recovery
Labor	91.5	112.0	81.7	$(3,307)	$3,511	$(6,818)
Materials	88.3	97.9	90.3	(3,099)	(478)	(2,621)
Energy	87.8	113.6	77.3	(460)	367	(827)
Capital	106.4	100.7	107.7	2,196	261	1,935
Total	95.5	104.2	91.7	$(4,670)	$3,661	$(8,331)

The first three columns provide indexes of productivity, price recovery, and profitability (productivity times price recovery) for each of the major inputs and in total. Examining the "Total" line, one can readily conclude that a 4.5 percent decline in profits (100 − 95.5) resulted from a substantial decline in price recovery (the company was not able to pass through increases in input costs to the customer) partially offset by a 4.2 percent increase in total productivity. The contribution of each of the inputs to this overall performance is readily discernible in the body of the report. In essence, the left side of the report provides us with partial and total measures of both productivity and price recovery.

The last three columns of the report provide the dollar impact of the changes in the indexes. This information provides perspective—a large percentage decline in the productivity of a minor input may be less serious than a smaller percentage decline for a major input. It also provides a tie-in to the income statement, enabling us to analyze profit changes in terms of their productivity and price recovery components. We may now determine objectively if we are a company "A" or company "B" (Chapter 1).

The mechanics of this particular measurement system are detailed in Appendix B.

A measurement system such as that described above enables management to quickly grasp the productivity performance of the company and its major operating components. It also greatly strengthens the analytical power of the planning process; the long-range impact of various assumptions regarding productivity and price recovery can easily be discerned.

MEASURING WHITE COLLAR PRODUCTIVITY

An organization's white collar activities, particularly those involving knowledge workers, require special consideration from a measurement standpoint. While the productivity of some white collar groups—particularly those whose work is repetitive or clerical in nature—is susceptible to measurement through the output-divided-by-input approach, others simply cannot be dealt with so cleanly and simply. The productive processes of many white collar functions are much less tangible than those of manufacturing, and outputs may be nearly impossible to define, much less count.

The very conception of productivity changes in a white collar environment. Even if outputs could be defined, maximizing the quantity produced relative to the inputs utilized may not be appropriate as the key performance objective. Which performance outcome is more desirable for a research and development organization—more new products, or products of greater marketplace value developed in a more timely fashion? What should be the major objective of a management information systems group—to develop as many systems as possible with its given resources, or to meet its internal customer needs as effectively as possible?

To be sure, increased efficiency in producing outputs is an appropriate objective for virtually any white collar organization. But in most cases, there will be other objectives that are of equal, if not greater, importance. Quality, in the sense of providing a service that is free of error, is invariably an important factor. Timeliness is also a significant performance variable in many white collar functions, and value to the customer, be it internal or external, must certainly be a high priority.

While white collar performance seems to defy definitive quantification, a *family of measures* can be a useful tool for monitoring the effectiveness of a white collar group. Under this approach, several measures, typically four to six in number, are selected and aggregated to provide an indicator of organizational effectiveness. The measures selected are those that are deemed to most closely capture the major aspects of performance of the group in question. Weights can be applied to the individual measures to reflect management's view of their relative importance. An example of a family of measures system in use by a computer center is presented in Table 6-5.

Table 6-5
Family of Measures

Measure	Weight	Index	
		Period 1	Period 2
Downtime	.3	1.14	1.21
% Time User Deadline Met	.3	1.07	1.09
Rerun Time	.2	.97	.95
# User Complaints	.1	1.02	1.05
On-Line Response Time	.1	.91	.87
Total	1.0	1.05	1.07

In effect, the nature of performance is defined by the family of measures, with the aggregate number representing an indicator of overall effectiveness. The family will normally include measures that represent a variety of variables—efficiency, quality, timeliness, value to customer. It may even include measures of organizational climate or indicators of innovativeness. Through the use of weights, management can fine-tune the system in order to correctly reflect the relative priorities of the different variables.

While a family of measures can obviously be developed by management, a more effective approach is to enlist the participation of the employees of the organizational unit being measured. A particularly effective tool to accomplish this is the Nominal Group Technique (NGT).1 NGT is a group brainstorming process, but highly structured and controlled. It typically utilizes a vertical slice of the organization being measured in order to obtain a wide representation of viewpoints. Use of the Nominal Group Technique (or other participative approach) provides two important benefits: (1) it taps into the knowledge and expertise of the people doing the job, and (2) it earns employee commitment to the resulting measurement system. The latter benefit can be a particularly difficult one to realize in white collar measurement, as professional and technical workers often hold the opinion that their work is not susceptible to measurement. The Nominal Group Technique is discussed in greater detail in Chapter 8.

In general, the term productivity has a broader connotation in the white collar environment than in the factory. Efficiency in producing an output is only one aspect of the productivity issue for engineers, accountants, and the like. Of perhaps greater importance is the degree to which a staff group is providing services that are of value to the organization. It would be difficult to conclude that a white collar group was productive if it were providing services that were of little value to the organization, no matter how efficiently it was producing those services. This suggests that white collar measurement, rather than being limited to considerations of internal efficiency, should instead provide broader indicators of organizational effectiveness.

White collar productivity improvement will be discussed in greater detail in Chapter 12.

THE OBJECTIVES MATRIX

A popular variation of the family of measures approach is the objectives matrix, developed by the Oregon Productivity Center.[2] The objectives matrix, a sample of which is presented in Figure 6-1, enables the family of measures to be integrated into the organizational goal-setting practices.

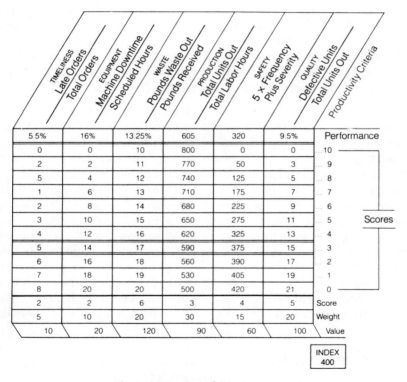

Figure 6-1. An objectives matrix.

The starting point is the family of measures, which are arrayed across the top of the matrix. The step-by-step process for constructing the matrix is as follows:

1. Enter the current level of performance for each measure on line 3 of the matrix.
2. Enter the goal for each measure in line 10 of the matrix.
3. Enter the minimum likely performance level for each measure in line 0.
4. Fill in the remaining lines of the matrix, either on a linear scale or in a nonlinear fashion. If, for example, it becomes progressively more difficult to improve as you get closer to the goal, it may be appropriate to utilize smaller increments for the higher numbers in the matrix.
5. Enter the selected weights in the appropriate line near the bottom of the matrix. As in the simpler family of measures, these weights are reflective of management priorities.
6. At the conclusion of each period, enter the actual value achieved for each measure in the row labeled "performance."
7. Circle or shade the value in the body of the matrix that represents the actual level of achievement for each measure; write the appropriate row number in the "score" row.
8. Multiply each measure's score by the appropriate weight, enter the result in the "value" column, and add the scores together to determine the overall result, which appears in the "index" box.

While somewhat more complex than a simple family of measures, the objectives matrix does provide a tie-in to a management-by-objectives system or some other goal-setting mechanism. It also provides for more flexibility in calculating the measures; increases in values of the measures do not have to be directly proportional to increases in the variable being measured.

Companies that have made extensive use of the objectives matrix include Northern Telecom and Boise Cascade. Northern Telecom, in fact, issued a 15-page booklet entitled "Managing With Productivity Indexing." More than just a guide to the mechanics of the objectives matrix, the booklet expounds on the value of measurement and even delivers some basic productivity awareness, such as this definition of productivity: "At Northern Telecom productivity is the quality, timeliness, and cost effectiveness with which our organization achieves its mission." The booklet is an excellent example of one organization's attempt to utilize measurement as an awareness-building mechanism.

DRIVING WHITE COLLAR PRODUCTIVITY THROUGH MEASUREMENT

Those who are skeptical about the effectiveness of measurement in a white collar environment need only look at the example provided by Southern Company Services, a subsidiary of the Southern Company.[3] The Southern Company is a holding company for four electric utilities in the Southeast: Georgia Power, Alabama Power, Florida Power, and Gulf Power. Southern Company Services (SCS) provides engineering, technical, and other services to the operating companies.

The Engineering Division of Southern Company Services employs 1,350 people in several functional groups. Activities range from the design of nuclear power plants to drafting, procurement, quality assurance, and other functions to support the construction of new plants.

Despite widespread employee skepticism about the feasibility of measuring engineering productivity, top management of SCS Engineering chose to address the measurement issue head-on as an initial step of the improvement effort. This decision was made in recognition of measurement's value as a communications, goal-setting, assessment, and accountability tool.

One of the division's engineers, David Connell, was appointed to research the productivity issues and to head up the effort.

The first step taken by Connell was to develop a meaningful definition: "Productivity is the relationship of how well an organization utilizes and converts its resources (manpower, material, equipment, capital, and energy) through some type of production process into company outputs (tangible items or services)."

Managers were then asked to develop measures that were consistent with the definition. To assist in that process, it was suggested by Connell in a series of awareness presentations that measures might fall into one of several categories:

☐ Effectiveness—producing the desired results.
☐ Efficiency—producing a minimum of waste, expense or effort while holding quantity and quality constant.
☐ Quality—a degree of acceptability as defined by standards or objectives.
☐ Quantity—measuring countable output in terms of increasing or decreasing magnitude.
☐ Timeliness—produced or provided on an accepted schedule.

Of greater significance than the measures themselves is the use to which they were put. The measures selected by each functional group, after approval by the division head, were graphically represented by a professional illustrator, framed, and displayed on the walls of a conference room built especially for that purpose. The room was dubbed the EPIC room—the Engineering Productivity and Information Center. Approximately 100 charts effectively covered the four walls of the room.

The EPIC room provided an excellent means of providing visibility for the measures. The charts were updated monthly, and all division employees were encouraged to enter the room and peruse the charts at any time. The head of the Engineering Division often consulted the measures himself in assessing managerial performance and in making headcount decisions. If, for example, the workload for a given area was declining, a negative trend would likely become apparent in the relevant measures in the EPIC room. This phenomenon generally led to workforce reductions through normal attrition.

In general, the EPIC room became a focal point for performance improvement in engineering. A competitive spirit developed as employees focused on improving the measures. Many units posted copies of their respective charts in their own work areas. While the measures certainly were not precise, they did serve as valid indicators of organizational performance and effectively drove the productivity improvement effort.

NATIONAL AND INDUSTRY DATA

The complete productivity measurement system will also make use of the "macro" measures of national and industry productivity. These data are primarily useful in gauging the organization's productivity performance relative to others in industry.

The U. S. Bureau of Labor Statistics (BLS) regularly publishes labor productivity data for the economy as a whole and for the manufacturing sector. BLS also publishes an annual report of labor productivity trends in 85 selected industries. The U.S. Department of Commerce, in its annual "Census of Manufacturing," provides data for all manufacturing industries that can be used to calculate productivity ratios. The Commerce Department's annual "Industrial Outlook" also provides productivity data for broad manufacturing and service industries. These data are potentially useful for gaining at least a rough sense of how the firm is performing relative to others in its industry.

Potentially the most useful macro-data are those obtained through an interfirm comparison study. These projects are typically sponsored and administered by an industry's trade association, which defines measures that are relevant to its industry, collects data from its members to calculate the values of these measures, and reports back to each member organization its relative position in the industry.

One of the most successful endeavors of this kind was that conducted by the American Productivity Center for the baking industry. Sponsored by the American Bakers Association (ABA) and the Bakery Equipment Manufacturing Association, this study compared data from approximately 200 participating bakeries and bakery chains. Working with the sponsoring organizations, APC developed relevant productivity measures, both physical and financial, for three major functional activities of bakeries: production, route sales, and thrift stores. Examples of some of these measures are presented in Table 6-6.

Table 6-6
Sample Productivity Measures
Bakery Interfirm Comparison

	Physical Measures	Financial Measures
Production:	Lb Out/Direct hour	$ Out/$ Total Labor
	Lb Out/Lb Flour	$ Out/$ Packaging
	Lb Out/Oven Capacity	$ Out/$ Maintenance
Route Sales:	Lb/Salesman	$ Out/$ Salesman Payroll
	Lb/Point of Sales	$ Out/$ Accounts Receivable
	Lb/Gallon Fuel	$ Out/$ Vehicle Cost
Thrift Stores:	Lb/Employee Hour	$ Out/$ Labor Cost
	Lb/Square Foot	$ Out/Customer
	Route Lb/Thrift Lb	$ Out/$ Assets Employed

Questionnaires requesting the data required to calculate these measures were mailed directly to ABA members by the American Productivity Center. The questionnaires were coded with a confidential identification number, which was known only to APC and was the only direct identification. Returned data were processed by computer, and reports were prepared showing the decile-by-decile industry performance for each of the productivity ratios. Each participating bakery received a report showing its own data and its decile location for each of the indicators.

The initial survey obtained and published data for the first half of 1982. The survey was again conducted for the first half of 1983, thus enabling the calculation of 1983-over-1982 trend data as well as the absolute levels of the indicators.

An interfirm comparison must, of course, be conducted by a neutral organization that can be trusted to maintain the confidentiality of sensitive competitive data. If such a study can be organized and successfully executed, the resulting data can be invaluable to participating firms in gauging their relative position in their industry.

MEASUREMENT IN PERSPECTIVE

The ideal measurement system will provide a means of tracking the productivity of every major organizational component and of every significant input. The measures will be tailored to fit the organization and the uses for which they are intended.

While measurement is integral to the productivity management process, it must be stressed that measurement is a tool and not an end in itself. The perfect measurement system, one that is not susceptible to any distortion whatsoever, is probably impossible to achieve. What is important is that some basic decisions about measurement be made early in the productivity planning process and that some reasonably reliable measures be devised to provide an indication of trend. As the productivity management process unfolds and develops, the measurement system can be expanded and refined in a deliberate and organized fashion.

REFERENCES

1. Delbecq, Andre L.; Van de Ven, Andrew H.; and Gustafson, David H., *Group Techniques for Program Planning*. Glenview, IL: Scott Foresman and Company, 1975.
2. Riggs, James L. and Felix, Glenn H., *Productivity by Objectives*. Englewood Cliffs, NJ: Prentice-Hall, 1983.
3. *Case Study 31: Southern Company Services*. Houston, TX: American Productivity Center, 1984.
4. Phillips, J., *Handbook of Training and Evaluation Measurement Methods*. Houston, TX: Gulf Publishing Company, 1983.

7

Human Resource Strategies for Productivity

THE ROLE OF THE HUMAN RESOURCE

An effective productivity management process is simply not feasible without the commitment and involvement of employees at all levels. How, after all, can an organization expect to create a performance-oriented culture when a large proportion of the workforce is neither committed to, nor involved in, productivity improvement?

As was noted in Chapter 2, productivity "programs" often pay scant attention to the potential contributions of the work force at large, relying instead on management employees to achieve higher levels of productivity. Since the involvement of the workforce is minimal, the results are predictable: apathy, mistrust, and resistance. Productivity improvement comes hard, as management must drag an unresponsive and ponderous organization to higher levels of performance against its will. The very concept of a productivity management process is not meaningful in this type of environment.

Those organizations that recognize the role of the human resource in productivity improvement and appreciate the power of a committed and involved work force typically devote substantial resources and management energies toward the development of an environment in which employees can and will contribute to performance improvement to the maximum of their capabilities. These efforts are typically characterized as quality of worklife (QWL) or employee involvement strategies.

WHAT IS QWL?

The quality of work life movement has grown apace in recent years, but its dimensions and its implications are often not fully appreciated. QWL is unfamiliar ground to many managers, and as a result, many are uncomfortable with the concept.

Most formal definitions of QWL describe a congruence between personal and organizational goals, i.e., ". . . the degree to which members of a work organization are able to satisfy important personal needs through their experiences with the organization."[1]

In a less rigorous vein, D. L. "Dutch" Landen, the former director of organizational research and development for General Motors, describes a high-QWL environment as one in which people are ". . . essential members of an organization that challenges the human spirit, that inspires personal growth and development, and that gets things done. . . ."

An interactive approach often used to assist managers and employees to relate to QWL in a practical way is called the "best/worst exercise." A group of people is asked to think back for a moment and recall the best organization with which they have ever been associated. Candidates are not limited to business organizations, but can include social or church groups, athletic teams, or other institutions. Having identified the best organization, participants are asked to write down words and phrases that describe *why* that organization was the best. They are then asked to do the same for the worst organization with which they have ever been associated. The words and phrases for both the best and worst organization are reported out by the group and posted.

The results are predictable and show little variation from one group to the next; typical responses are presented in Figure 7-1. What is significant about this exercise is the common thread that runs through virtually all of the responses: the people orientation. Rarely are phrases such as "financial strength" or "market leadership" or "leading-edge technology" nominated

---

Best Organization

Good leadership	Teamwork/cooperation
Common goals	Pride in company
Recognition	Good communication
Participation	Opportunity for growth

Worst Organization

Poor leadership	Lack of appreciation
No involvement	No direction
Poor communication	Low productivity
Low morale	Excessive pressure

Figure 7-1. Typical best/worst organization responses.

for the lists. Rather, when asked to characterize the best organization, people tend to think in terms of the environment within which members of the organization functioned. Reflection on the qualities of the best and worst organizations tends to bring QWL into sharper focus.

Interestingly, the best/worst exercise yields strikingly similar results for every level in the organizational hierarchy, from top management, through middle management and first-line supervision, to hourly labor. This suggests that QWL does not mean radically different things at different levels of the organization.

We can define QWL in terms of some of the conditions that characterize a high-QWL environment:

Employee input to decisions. When employees have the opportunity to influence decisions that affect their jobs, they feel more in control of their destinies and less like meaningless cogs in a giant machine. Allowing employees' input to decisions does not mean surrendering management prerogative; decision-making responsibility remains vested in the manager. It does, however, mean careful consideration and honest evaluation of that input.

Employee participation in problem-solving. The active solicitation of employees' ideas and their involvement in problem-solving processes greatly enhances the ability of employees to make a contribution to the organization and to derive a sense of accomplishment and value.

Information sharing. Individuals have a greater sense of belonging to an organization when they are fully informed of developments and forces shaping that organization. Handing employees copies of the company's annual and quarterly reports isn't sufficient; information sharing means continually informing employees about developments impacting the organization and explaining to employees the reasons behind management decisions. Informed employees are also better able to solve problems and react positively to changing conditions.

Constructive feedback. The need to know how we are doing and how we might improve our performance is a characteristic of human nature. Frequent and regular feedback reinforces good performance and encourages self-development.

Teamwork and collaboration. Most people enjoy being part of a team and working with other members of the organization toward the achievement of shared objectives. An awareness of and commitment to the company's mission and objectives provide employees with a sense of direction and common goals.

Meaningful and challenging work. Employees who have responsibility for a range of tasks and enjoy a degree of autonomy and discretion can be expected to derive greater satisfaction from their work than those whose jobs are repetitive and boring.

Employment security. It is difficult to convince an individual that he enjoys a high quality of work life when he fears for his job. In any effort to improve productivity, employment security invariably becomes an issue with employees, who wonder if, through contributing to the improvement process, they will be eliminating fellow workers' (or their own) jobs. As suggested in Chapter 3, a powerful means of overcoming this barrier, while simultaneously improving quality of work life, is to commit to employees that no jobs will be lost as a result of productivity improvement.

EMPLOYEE INVOLVEMENT

A major tenet of quality of work life, and the common denominator of most QWL efforts, is employee involvement. Employee involvement programs are now widespread in American organizations and offer great potential for fostering organizational change.

Unfortunately, many employee involvement efforts suffer from a fatal flaw: they are viewed by management as simply another technique to bring about productivity improvement.

Those organizations that approach employee involvement only as an improvement technique tend to focus their efforts on popular involvement structures, such as quality circles (see Chapter 8), with the objective of proliferating their use in ever greater numbers throughout the organization. Techniques, however, tend not to last; their effectiveness in improving performance plateaus and organizational enthusiasm ultimately wanes. In addition, traditional organizational practices and systems may not be supportive of these techniques, which further contributes to their ultimate demise.

The real power of employee involvement lies in its ability to bring about cultural change by fostering a more participative management style in an organization. Participative management does not equate to quality circles or any other specific involvement technique. Rather, as Dutch Landen puts it, ". . . participative management is an organized set of ideas about *how* to manage a modern and progressive business enterprise."[2]

While a participatively-managed organization certainly utilizes techniques like quality circles to involve employees in problem solving, involvement goes much deeper. Employees participate in goal setting, for example, and in planning activities. Employees are fully informed about organizational direction, strategies, and performance. They have input to decisions that affect them, and their skills are developed in order that they may grow and make ever greater contributions to organizational performance. They are cognizant of the need to adapt to a changing environment and are involved in bringing about that change.

ASSUMPTIONS UNDERLYING INVOLVEMENT

It should also be recognized that there are different degrees of involvement, with varying assumptions supporting them. Lawler has divided participative approaches into three categories, based on their underlying assumptions.[3]

The *human relations* approach is based upon the belief that more satisfied employees are more productive employees. While tapping into employee ideas may provide some direct performance improvements, the primary motivation for an involvement initiative under this assumption is to raise the level of job satisfaction and reduce employee resistance to change. Organizations operating under the human relations framework utilize involvement techniques such as suggestion programs, survey feedback, and quality circles (Chapter 8), but do not seek to redesign the jobs, change the organizational structure, or transform the organization's culture to promote maximum employee involvement in decision making. Perhaps the majority of employee involvement initiatives in the United States are based, either explicitly or implicitly, on the human relations assumptions.

The *human resources* approach goes beyond the human relations approach. The key assumption here is that people are a valuable resource, capable of making significant contributions to organizational performance. They should be developed in order to increase their capabilities, and when they have input to decisions, better decisions result.

Organizations operating under the human resources assumptions are more likely to invest in a long-term change process to develop more participative practices on the job and to foster the notion that the supervisor and his subordinates are a team. They move beyond quality circles and attempt to integrate involvement into the routine, day-to-day functioning of the organization. They may implement some limited job redesign and often install gain sharing systems (Chapter 9).

The *high involvement* approach carries participative management even beyond the human resources model. High involvement systems operate under the assumption that employees are capable of making important decisions about their work and that maximum organizational performance results when people exercise considerable control over their work activities.

High involvement organizations typically utilize profoundly different approaches to job design, such as autonomous work teams (Chapter 8). They are very flat organizations, as employees make most of the routine, day-to-day decisions that are made by supervisors in traditionally managed organizations. All of the organization's systems, such as the reward system and the goal-setting system, are designed to reinforce maximum employee involvement in decision-making. High involvement systems represent a radical departure from traditional management assumptions and thus require a great deal of management commitment to change.

Lawler rightly emphasizes the need for management to clearly define and articulate the assumptions that underlie their employee involvement initiative. If the organization's approach to participation is not congruent with these assumptions, failure will surely result. In addition, if these assumptions are not well understood and accepted by other members of the organization, support and buy-in will be lacking, and management's vision of employee involvement will not be effectively operationalized.

WHY PARTICIPATIVE MANAGEMENT NOW?

A history of success can be a burden if it inhibits change. Our nation's extraordinary economic and industrial success leads many managers to question the need to change. After all, if our management methods have resulted in decades of continuous success, why should we change our style now? If we have problems, why can't we just work our way out of them like we have always done?

Our traditional style of management has its roots firmly embedded in the scientific management principles espoused by Frederick Taylor in the early 1900s. The beliefs and philosophies underlying this style may be summarized as follows:

☐ The technological imperative. Technology is viewed as the prime determinant of organizational performance, and plants are designed to maximize the effectiveness of the technical system. The role of people in these organizations is not considered until after the technical system has been designed, and this role is therefore completely dependent upon the technical system.
☐ People are extensions of machines and expendable spare parts. The primacy of technology relegates people to a secondary role, a necessary evil required to run the machines.
☐ Maximum task breakdown. Organizational tasks are subdivided to the greatest degree possible, so that each employee's job is highly prescribed and very repetitive. The job is made "idiot proof," and employees are not allowed to exercise discretion in the way that they execute their responsibilities.
☐ Simple, narrow skills. Because jobs are designed to be simple and repetitive, employees require only minimal skills.

☐ Reliance on external controls. Management ensures that work is being done as it was designed through extensive use of controls—work standards, procedures, and policies.
☐ Tall organization chart. The control-oriented environment requires large specialist staffs to develop controls, set standards, write procedures, and monitor compliance.
☐ Quantitative orientation. Performance against standards is monitored closely, efficiency improvements are emphasized, and quantitative analysis is extensive.
☐ Autocratic management. The authoritarian, command-style of management fits nicely with the other characteristics. Employees are held in low esteem and are deemed to have little to contribute beyond the execution of their controlled, repetitive jobs. It is management's job to make decisions, solve problems, and direct employees.

This style of management was successful in the United States for many decades and continues to be the norm. It worked fine during the late forties, fifties, and early sixties, when we enjoyed unparalleled economic growth and expanding world markets. The key to success in this era was to organize our resources to produce products in as great a volume as possible to meet the ever-growing and seemingly insatiable demand. Cost wasn't terribly important—just pass it through to the customer.

But times have changed. No longer can we count on continuing economic growth; economic cycles seem to have become more pronounced and unpredictable. Even in times of economic growth, as in the mid-eighties, all of our industries do not participate. Even more important, we are in an era of global competition. The Japanese have attacked numerous American industries with great success. The South Koreans are exporting televisions, computers, and automobiles to the United States. Even the Yugoslavians have invaded the American auto industry with a low-priced car. Competition is everywhere, as country after country takes aim at lucrative American markets.

Aggravating the impact of economic and business changes is a perceptible change in the values of our workers. The worker of the 1950s typically made a lifetime commitment to his company. His orientation was toward providing for his family and building assets for his retirement. He was loyal to his employer and did not expect to have much influence over how his job was done. Today's worker, by contrast, is a different sort. His orientation seems to be more short-term. He is better educated and is not satisfied to spend one-third of his life in a job that provides no satisfaction or psychic rewards. He demands influence over his life at work.

Our traditional management style is not as effective in an environment of turbulent economic change, intense competition, and today's values. Indeed, it was designed for another era.

Corporate America has begun to recognize the shortcomings of our traditional management style, and efforts to transform corporate styles and cultures have begun. Present management practices are giving way to a new management philosophy that recognizes that people are the key to success in any organizational endeavor.

The philosophical underpinnings of the new management style are very different from those of the old:

Sociotechnical optimization. No longer are savvy companies attempting to optimize the technical system in designing new organizations. They now recognize that the social system—the way that people fit into the organization—is also an important determinant of performance. Unfortunately, the maximization of the technical system usually results in a poor social system; jobs are boring, teamwork is nonexistent, and people have little influence over their lives at work. Experience has shown, however, that by consciously designing a more effective social system, even at the expense of the technical system, better organizational performance results. So organizations are now being designed to provide for an optimum fit between the technical and social systems.

People are complementary to machines and a resource to be developed. People can make significant contributions and can enhance the effectiveness of the technology if jobs are designed to challenge employees and to provide them with opportunities to exercise their abilities and gain additional skills.

Optimum task grouping. Rather than subdividing tasks into their smallest possible elements, companies are now expanding and enriching jobs in order to provide employees with greater responsibilities, knowledge, and challenge.

Multiple, broad skills. To maximize the contribution of the human resource and to enable people to effectively execute expanded responsibilities, extensive training is provided to broaden and expand employee capabilities.

Reliance on self-regulation. Extensive external controls are giving way to greater self-management based on a shared vision and values. Employees are given greater influence over how their work is accomplished and are provided information and support with which to make informed decisions.

Flat organization chart. The increased use of self-management enables supervisors to exercise a broader span of control, and the lessened reliance on controls reduces the need for specialist staffs.

Qualitative orientation. Product quality, process quality, and quality of worklife are watchwords of the new management style. Organizations are driven by concerns of quality, innovation, and value rather than by quantitative analysis and efficiency.

Participative management. The new management philosophy, which emphasizes the commitment and involvement of the human resource, requires management practices that support these ends. The autocratic management style would be totally inappropriate in this environment.

Participative management, then, is simply a manifestation of organizations adapting to a changed environment. The traditional style, which worked fine in an era of stability and steady growth, is ineffective in a period of intense competition and rapid change. Maximum quality, productivity, and responsiveness to change are keys to success in this era, and an informed, committed, and able workforce is a prerequisite for the achievement of these outcomes.

Our business environment is very different from that of 20 years ago. How can we expect to succeed with a management philosophy that was designed in a different era under very different conditions? History has taught us repeatedly that organizations that doggedly maintain their traditional ways of doing things, rather than adapting to changes in their environment, do not survive.

QWL, EMPLOYEE INVOLVEMENT, AND PRODUCTIVITY

How do quality of work life and employee involvement relate to productivity? To answer that question, one needs only to refer back to the best/worst exercise for guidance. Can there be any question whether the best or the worst organization has the more committed employees? Is there any doubt about which organization, external and marketplace conditions being equal, will be the more successful?

An organization that attends to the quality of its employees' work life will reap the benefits of a more committed work force. The manifestations of a committed work force include a closer identification with the organization, a willingness to cooperate with management in improving performance, and a desire to see the organization succeed. A more committed work force is a more productive workforce.

In addition, the employee involvement aspect of QWL results in better decisions by tapping into a tremendously under-utilized resource: the minds of employees. The collective thoughts and ideas of employees, when channeled toward improving organizational performance, represent a veritable gold mine of opportunity. We may escape without fully utilizing this resource if we enjoy rapid market growth and low competition. But in times of turbulent environmental change and intense competition, it is imperative that we maximize each employee's contribution to organizational performance.

ARTICULATING THE NEW PHILOSOPHY

One of the major companies undertaking a change in its fundamental management style is Motorola. This company provides an excellent example of an effort to articulate and communicate this change to the organization. In a slick, eleven-page brochure, Motorola itemizes its assumptions regarding people:

- ☐ Every worker knows the job better than anyone else.
- ☐ Employees can and will accept the responsibility of participating in the management of their work if that responsibility is presented to them properly.
- ☐ The sum of individual commitment and shared ideas is greater than the individual contributions made by managers.
- ☐ Intelligence, perspective, and creativity exist in the same proportion among people at all levels of the organization.
- ☐ We, as managers, have been trained to put demands upon the work force; we, as workers, have been trained to respond to those demands; we, as participants, must enter a dialogue with one another and learn to work together.

The brochure goes on to define participative management at Motorola:

- ☐ A structured, yet flexible way of managing the company on a continuing basis.
- ☐ A management system, not a plug-in program.
- ☐ Managerial encouragement and support of teamwork, idea-sharing, and mutual trust.
- ☐ Increased two-way communications about the goals and objectives of the business, about how to achieve them, and about specific progress toward them.
- ☐ More decision-making at the most appropriate level (which is usually lower in the organization).
- ☐ More employee responsibility—where direction, discipline, and control are generally self-motivated, and not externally imposed.
- ☐ A financial sharing of the benefits of improved productivity, quality and service, between the employees and the company.

DEADLY MISCONCEPTIONS

Employee involvement efforts often flounder in a sea of misconceptions. Any attempt to change an organization's fundamental style of management will be difficult under the best of circumstances. But when managers misunderstand the very nature of participative management, the task is well nigh impossible.

Unfortunately, misconceptions about employee involvement abound. Some of the more common ones are described below:

Loss of prerogative. Perhaps the most common misconception is that managers lose their prerogative and no longer make decisions. This perception exists at all levels within organizations; a vice president of a utility, after hearing a presentation on employee involvement, threw up his hands and remarked with great sarcasm, "Why don't we just turn the company over to the employees?" This interpretation could not be any less correct. Employee involvement is not an abdication of management responsibility, it is rather a more disciplined form of management. Managers still make decisions, but they make full use of all the resources at their disposal in making those decisions. When the ideas and capabilities of employees are utilized fully, better management decisions result.

Industrial democracy. Characterizing employee involvement as "democracy" is an unfortunate use of the word, for it implies that decisions are made by popular vote. Any manager that puts every decision to a vote of his subordinates is not managing participatively; he is not managing at all.

Coddling employees. Those who view employee involvement as being "soft" on employees adhere to the macho theory of management. The manager is a tough guy who barks orders and doesn't brook any back talk from his lazy and irresponsible employees. This orientation reflects a very negative and cynical view of human nature and the contributions that can be made by the human resource.

A technique for the bottom of the organization. The assumption is that employee involvement only applies to the lower levels of the organization, to nonmanagement employees. Employee involvement is something that management does to hourly employees. Holders of this viewpoint assume either that supervisors and managers are already managed in a participative fashion, or do not need to be. They usually are not, and expecting people to manage in different ways than they themselves are managed is expecting too much.

KEYS TO SUCCESS

Success in the employee involvement arena requires, first and foremost, a recognition by top management that participative management means *cultural change*. The bulk of today's supervisors and managers earned their spurs in the days when the traditional management style was the norm. They have always gotten things done through the use of authority. They are autocratic managers.

Other management systems and practices support the traditional management style. Goals are set at the top, without input from most employees. Communications are limited, secretive, or virtually nonexistent. Recognition practices and systems are not well developed. Labor relations are poor, and the union-management relationship is highly adversarial. Hourly employees are promoted to supervision based on their technical skills rather than their people-management capabilities. Little formal training is provided for managers, and virtually none is available to hourly employees. Collaboration and teamwork are minimal.

It is little wonder that employee involvement cannot thrive within today's typical organizational environment. If employee involvement is to succeed in the long run, that environment, that culture, must change. Ingrained attitudes and entrenched systems must change. Otherwise employee involvement will have no more success or lasting impact than any of the past techniques that have come and gone. It will ultimately be rejected by an organization whose systems and practices are unsupportive.

If employee involvement is viewed as a cultural change process, its implementation will be approached in a very different manner than if it is viewed as an improvement technique. An appropriate implementation strategy will be presented later in this chapter.

Cultural change, of course, requires *management commitment* and a *long-term perspective*. People resist change, as it requires behaviors and responses that are unfamiliar. The rationale for change may not be clear, and people may feel unable to cope with the new requirements implied by change.

Management commitment to change must be apparent and unambiguous if this resistance is to be overcome. Management must be willing to support change through the provision of resources, the modification of organizational systems, and personal involvement in the change process.

Management must also adopt a long-term view if change is to succeed. Attitudes and behaviors do not change overnight, and management demands for quick success will heighten resistance and undermine the process.

Another key to success in employee involvement is *supervisory support*. The person who generally feels the greatest pressure in an employee involvement effort is the first-line supervisor. He is the buffer between top management and nonmanagement employees, and it is upon him that the greatest pressures fall. He is called upon to transform his familiar and comfortable style, yet he lacks the knowledge and skills to do so. What's more, he may well disagree with the very concept of employee involvement. He feels very insecure and very threatened.

If the first-line supervisor's needs are not attended to, there is a strong likelihood that he will resist. His resistance will be subtle, rather than open, because he does, after all, wish to keep his job. He will go through the motions, but he will not change.

Since supervisory support is such an essential ingredient to this process, it is imperative that we take steps to obtain that support. For starters, supervisors (and middle managers) must be educated about employee involvement; they must understand what it is, why we need it, and how they will be supported. They must appreciate the fact that this is not just another top-management-initiated program to wring more gains out of the organization, but is rather an effort to adapt to a changed environment. They must realize that their effectiveness as supervisors will be enhanced by this initiative.

The resistance of supervisors can be further reduced by involving them in planning and managing the employee involvement process. If we recognize that one of the major benefits of employee involvement is increased commitment and ownership, it is only logical that we apply these principles to the supervisors themselves. Their input to the plan and its ongoing execution will provide them with a sense of ownership and control that could not be obtained otherwise. In addition, their involvement will probably result in a better plan.

We must also provide the supervisor with the skills he needs to manage participatively. To be effective, he will need skills in group leadership, active listening, communications, providing feedback, and problem-solving. Training, coaching, and reinforcement in these skills is essential.

Finally, supervisors need to be given some relief from other priorities if they are to be expected to concentrate on involving employees. Time requirements are particularly heavy in the early stages of an employee involvement effort, and a barrage of high-priority, short-deadline requirements will simply ensure that supervisors will be unable to devote the requisite time and effort to employee involvement.

Another key to success is *union support*. If the union as an institution is not involved in the employee involvement initiative, they may well resist the effort. And if they have any influence at all with employees, their resistance may cause the effort to fail. One plant learned about union resistance the hard way—union officers, standing outside the door of the meeting room where managers had invited employees to learn about the organization's employee involvement program, advised employees not to attend. The effort was materially damaged and ultimately failed.

Union resistance is a manifestation of the lack of trust between management and the union. If the union is left out of the effort, they will naturally assume that management has ulterior motives, such as eliminating jobs or undermining the union's influence with its members. Strategies to gain union support for employee involvement and productivity improvement are discussed in Chapter 11.

Another key ingredient of any employee involvement effort is *training*. The need for supervisory skills training was described earlier, but the need is much broader than that. Awareness training, for example, must be conducted at all levels of the organization. Managers and supervisors must appreciate that participative management represents a major change from traditional styles of management. If they are to be expected to embrace this change, they must understand the nature, rationale, and implications of participative management.

Training in problem solving must also be provided. The typical employee does not have well-developed skills to analyze problems and to develop solutions. If employees are to be engaged in problem solving, the requisite skills must be obtained if they are to be effective.

Finally, employee involvement requires a well developed *strategy* if long term success is to be achieved. More than any other element of an effective productivity management process, employee involvement challenges long-held beliefs and impacts broad areas of organizational functioning. Changing management style is probably the most difficult and frustrating task facing the executive who desires to institutionalize productivity improvement. At the same time, it may be the most important task. An intelligent, long-term strategy is therefore a vital ingredient for success. A framework for such a strategy is presented in the following section.

AN EMPLOYEE INVOLVEMENT STRATEGY

Many companies undertaking employee involvement efforts lack a change strategy. Their activities are strictly tactical, focusing upon the implementation of a specific involvement technique, such as quality circles. A consultant is hired, a coordinator is appointed, quality circle training is provided, and circles begin to sprout in every nook and cranny of the organization.

There is nothing inherently wrong with quality circles; they can be a very effective employee involvement technique. But they are only a technique. If the organization is not prepared, and the barriers to cultural change are not addressed, quality circles will not last and participative management will not be institutionalized.

Since employee involvement will likely represent the biggest change for the organization, an effective employee involvement change strategy is vitally important to the success of the effort. A change strategy will consist of several steps, such as the following.

Step 1: Management and Supervisory Awareness

Because participative management represents a considerable change from traditional management styles, it is imperative that management personnel at all levels fully understand the nature and rationale for the change. People do not willingly change their behaviors without a felt need and a degree of comfort with the new ways. Misconceptions about employee involvement are very widespread, and if not overcome, will almost certainly foster supervisory resistance.

Companies committed to a change process typically invest significant resources in management and supervisory awareness training. A number of organizations, for example, have conducted full-day employee involvement awareness programs for all of their supervisory and management staff (and nonmanagement employees as well in some cases). A typical agenda for an employee involvement awareness program follows:

☐ Overview of productivity: definition and recent U.S. and international trends
☐ Managing productivity: the role of the human resource
☐ Overview of employee involvement:
  • Definition—What is it?
  • Key elements of participative management
  • Requirements for success
  • Misconceptions about employee involvement
  • Barriers to employee involvement
☐ Experiential exercise: a group exercise to demonstrate the benefits of employee involvement
☐ Employee involvement techniques: programs of varying degrees of sophistication to bring about participation of employees at the working level
☐ The role of the union in employee involvement
☐ A change strategy for implementation of an employee involvement process
☐ Group exercise: what are the benefits of and barriers to employee involvement in this organization?

Step 2: Readiness Assessment and Strategic Planning

It should be apparent by now that employee involvement is a major undertaking with many potential barriers to success. It is important that the organization identify those barriers in order to evaluate its readiness to support a cultural change effort. There are a wide range of potential readiness problems: the level of trust between employees and management may be low; communications practices may not be sufficiently developed to support the process; a noncollaborative relationship between management and the union may cause union resistance; organizational awareness of employee involvement may be low; insufficient resources may be earmarked for the effort; and middle management commitment may be inadequate to support cultural change. The identification and recognition of these issues is vital, for if they are not addressed early in the process, failure may well result.

A readiness assessment is most effective if it involves employees at all levels; a cross section of employees may be interviewed, or broader input may be obtained through the use of a written survey. A simple survey instrument for evaluating readiness is presented in Appendix C.

Armed with a knowledge of the key readiness issues, management is now able to develop a strategic employee involvement plan. This will not be a detailed tactical plan, but will establish the general framework and direction for the effort. It will address the various readiness issues and will define in general terms how management expects to bring about the desired cultural change. A strategic plan is important to ensure that management style is changed through a deliberate, systematic, and integrated process.

Step 3: Tactical Assessment and Preparation

It is appropriate at this point to consider the tactical side of employee involvement. The purpose of the tactical assessment is to evaluate the wide range of employee involvement techniques in order to identify those that fit best with the organizational culture, style, and state of readiness. While employee involvement should not itself be approached as a technique, techniques are nonetheless required in order to foster and execute the change. We must provide systems and structures through which employees are able, in an orderly fashion, to become involved in problem solving and performance improvement. The problem is that there are many such techniques, some of which will be more appropriate and effective than others in any given organizational environment. By assessing and evaluating the various techniques, we are able to ensure that the effort is intelligently tailored to the existing circumstances. A discussion and evaluation of the various techniques is presented in Chapter 8.

The preparation aspect of this phase involves those activities that are designed to raise the level of organizational readiness for employee involvement. We ensure that the awareness of the various parts of the organization has been improved, that the support of the union has been enlisted, that communications practices have been improved, that needed resources have been made available, and that other barriers identified in the readiness assessment have been attended to.

Step 4: Training and Implementation

With awareness high and the organization in an improved state of readiness, we are now ready to provide the needed skills and to implement the techniques selected in the previous phase. Supervisors or team leaders are provided training in group leadership and other required skills, and employees are provided training in problem solving. Initial implementation efforts are probably limited to one or two pilot areas, where they can be heavily supported and carefully evaluated.

Step 5: Evaluation, Diffusion, and Maintenance

As with any major undertaking, we must continually evaluate progress, determine the degree to which our expectations are being met, and consider how we can do it better. To the extent that initial implementation efforts were confined to limited areas, diffusion throughout the organization must be accomplished. And the effort must be continually supported, maintained and reinvigorated.

SUPPORTING CHANGE

The lack of long term strategy to change management style has ensured the ultimate failure of many employee involvement efforts. The scenario described earlier—the hiring of a consultant to train employees and implement quality circles—is a manifestation of the failure to recognize employee involvement as a long-term change strategy. If we refer back to the implementation framework described in the previous section, it is apparent that the organization in this scenario has bypassed the first three phases, launching their employee involvement effort in phase 4—Training and implementation.

Bypassing the first three phases certainly saves time and money, but what are the consequences? Since no effort is made to build awareness, managers and supervisors have widely divergent perceptions of employee involvement. Most do not appreciate the need to change their management style and feel threatened by this new initiative. Since no readiness assessment is conducted, numerous organizational barriers that mitigate against the success of the effort are not attended to. The lack of a strategic plan leaves management without a clear sense of what is beyond quality circles and of what is the ultimate aim of this endeavor. Since a tactical assessment is not conducted, it is entirely possible that quality circles are completely inappropriate for this organization at this time, and that other involvement techniques that might have produced better results have been overlooked.

The predictable result is that the quality circle effort takes its place on the long list of company programs that showed promise, returned some short-term gains, but ultimately peaked and faded away. The traditional culture and management style remains intact, and it's back to business as usual.

The development of a participative management style to support a productivity management process represents a major change and thus must be supported by a change strategy.

REFERENCES

1. Hackman, J. Richard and Suttle, J. Lloyd, *Improving Life at Work*. Santa Monica, CA: Goodyear Publishing Co., 1977.
2. Landen, D. L., *Productivity Brief 16: The Future of Participative Management*. Houston, TX: American Productivity Center, 1982.
3. Lawler, Edward E., III, *High Involvement Management*. San Francisco, CA: Jossey-Bass Publishers, 1986.
4. Scobel, D., *Creative Worklife*. Houston, TX: Gulf Publishing Company, 1981.

8

Employee Involvement Techniques

TECHNIQUES IN PERSPECTIVE

The previous chapter warned against the tendency to equate popular techniques with employee involvement. As was suggested, employee involvement must be viewed as an organizational change effort, supported by management commitment, a long-term perspective, and reinforcing organizational systems.

Employee involvement techniques, nonetheless, do have their place. In fact, we cannot hope to achieve a change in management style without them. For techniques provide us with mechanisms to enable employees to become involved in a structured way. Ultimately, of course, we would like to reach a state where employees are involved in problem-solving and decision-making informally on a day-to-day basis. But until we get there, we must facilitate involvement through organized, structured mechanisms so that the organization can gain experience and skills in employee involvement.

Techniques, then, represent the *tactical* side of employee involvement. They are tools that we use in order to advance our greater strategic purpose—the development of a management style and a climate that maximizes the contributions of the human resource to productivity improvement.

Because they are vital elements of the change strategy, it is important that we utilize employee involvement techniques intelligently. There are a wide variety of involvement techniques, and they vary greatly in terms of their sophistication, organizational support requirements, and effectiveness in promoting change. We must not become so enamored of a particular technique that we fail to utilize other techniques that may be more appropriate. We must not attempt to utilize techniques that are not consistent with our present state of organizational readiness. And we must be prepared, when a given technique ceases to be effective, to institute new techniques in order to maintain the momentum of our effort.

THE TECHNIQUES CONTINUUM

Given the wide variety of employee involvement techniques, it is useful to organize them along a continuum such as the one presented in Figure 8-1.

At the left-hand end of the continuum are those techniques that are relatively simple, can be implemented without a major commitment of management time and resources, and can be reasonably effective in traditionally-managed organizations.

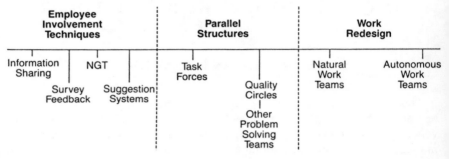

Figure 8-1. Employee involvement continuum.

The techniques at the far right-hand end of the continuum are another matter. These techniques represent the state-of-the-art in employee involvement. They are highly sophisticated, and because they represent a radical departure from traditional management practices, require extremely high levels of management commitment. They should not even be considered by an organization that is a novice in the employee involvement arena.

An appreciation of the techniques continuum is a valuable asset to any organization that desires to manage its employee involvement effort in a systematic and rational fashion. It enables the organization, for example, to select techniques that fit its present state of readiness and level of experience with employee involvement. In addition, it helps the organization maintain and constantly revitalize its effort by identifying techniques of ever-increasing sophistication and effectiveness.

Low-order Techniques for Involvement

At the beginning of our continuum, we find the low-order involvement techniques. These techniques have been around for a long time and have been utilized, to one degree or another, by most organizations. They do not require an exceptional amount of management commitment to be effective, and they need not consume an inordinate amount of resources. Designing and implementing these techniques is a relatively simple undertaking when compared with higher-order techniques. It should not be assumed, however, that careful attention to the design of these techniques is not important; a poor design will simply ensure that no benefit will be gained from their use.

Information Sharing

The most basic of all employee involvement techniques, information sharing should be a fundamental feature of all involvement efforts. Lacking information about organizational performance, the nature and challenges of the business, and the rationale behind decisions that affect their jobs, employees become alienated and resentful. They feel no loyalty toward the organization and are not committed to its objectives and endeavors. They are just another cog in the company's machine.

Managers often fail to appreciate the power of communications and information sharing. Good communications practices can mean the difference between commitment and indifference on the part of the work force. By keeping employees informed, management sends a powerful signal: It says that people are viewed as an important asset of the organization. Lack of communication sends an equally clear signal: Employees are not valued enough to be kept informed. One high-ranking executive was heard to comment, "We communicate with employees on a need-to-know basis." The attitude that it is not important to communicate any information beyond what is necessary to do the job is a manifestation of a very low regard for people.

Audrey Freedman, Chief Labor Economist for The Conference Board, said it well:

"Lack of information about company performance hampers employees, at all levels, from being efficient and effective members of an enterprise. Not knowing the cost of operation and competitive circumstances of their employer, workers misperceive the source of real wages. Moreover, having been shut out of knowledge about company performance, employees conclude that it does not in any way depend upon their actions. Thus, supervisory, managerial, and executive layerings create disinterest and conflict in an enterprise. All of this robs us of productivity and an engaged workforce.

"In the simplest terms, an employee does a more thorough and effective job when she/he knows the importance of the work, how it fits into the business, what exactly *is* the business, and how the employee's job future and wages are interdependent with the enterprise.

"Moreover, each layer of management is required to be more thorough and effective also—when business information is shared internally with all levels.

"Having inside information makes employees 'insiders'—which may seem like playing with fire. It is. It conveys a kind of ownership beyond what a stock dividend or profit-share provides. It engages the employee in common effort."[1]

Companies that value people and emphasize information-sharing regularly inform their work force about a wide variety of subjects:

☐ Organizational performance—sales, production levels, quality statistics, productivity data, budget performance, even financial data such as profits and returns. Many companies have concluded that the benefits of an informed and committed work force far exceed the risks associated with the communication of sensitive financial information.
☐ Nature of the business—types of products or services provided, use of products, nature of production processes, major customers, customer requirements, and major suppliers. Many manufacturing organizations send groups of hourly employees to customers' facilities in order to see their products in use and obtain feedback from the customers.
☐ Competitive circumstances—identification of major competitors and their strengths and weaknesses, nature of the marketplace, and competitors' strategies. An awareness of the competitive challenge can serve to motivate improved performance.
☐ Organizational initiatives and changes—rationale and nature of improvement programs, organizational changes, major capital investments, new technologies, product or service enhancements, and policy changes.

Organizations that have no prior experience with employee involvement would be well advised to begin by significantly improving their communications and information-sharing practices. Improved communications is a low-risk tactic and can serve to unfreeze the organization. And more than one organization has been surprised to find that improved communications returns business results. One plant, for example, attributed an increase in bottom-line performance to nothing more than a rigorous communications effort.

Formal communications efforts often rely on the first-line supervisor as a key disseminator of information. This person is typically ill-prepared for such a role, however, in part because he himself is not the recipient of good communications from above. The oft-heard phrase, "The supervisor is always the last to know," carries the ring of truth in many companies.

The Industrial Controls Division of Allen-Bradley met this problem head-on by designing a communications program that focused on preparing the first-line supervisor.[2] Dubbed *Communicating for Productivity and Survival* (CPS), the program is viewed by division management as a critical element of their productivity improvement strategy. The goal of CPS is to educate employees on the nature of the business, competitive challenges faced, and the division's strategic plans and objectives.

A central element of the program is the Interdepartmental Letter, or IDL. Prior to the appearance of bulletin board announcements and special communications events, an IDL containing in-depth information on the subject is delivered to all managers and supervisors. Prior to each of a series of short communications bulletins on Allen-Bradley competition, for example, supervisors received an IDL containing detailed information—history, products, strengths, strategy—on that particular competitor. The IDL is an effective tool to enable supervisors to answer questions and discuss the subject with employees at communications meetings.

Another interesting aspect of the Allen-Bradley program is the use of extensive structural entities to support the effort. At the top is a steering committee of senior and middle managers who choose the topics to be covered in CPS. An implementation committee of line managers and first-line supervisors reviews the content of the proposed communications and recommends changes to improve the understandability and acceptance of the message. Non-supervisory employees participate in the process through a 25-member implementation task force that provides a further review of the proposed communications. Finally, a facilitating committee of managers and supervisors coach other supervisors and obtain back-up materials for those desiring further information on the subject.

By all indications, the Allen-Bradley program is a success. Two years after the launch of the program, 95 percent of the hourly employees could identify all of the division's major competitors. A survey indicated that over 70 percent of employees felt that their supervisors' communications skills had improved over the course of two years, and small group meetings were rated an average 3.7 on a scale of five. In addition, the percentage of employees who viewed the supervisor as the prime source of information doubled from a previous survey.

Survey Feedback

Surveys can take a variety of forms. The written instrument is well known, but surveys can also be conducted through personal interviews or by observation. Surveys can also serve a variety of purposes. They can be used to gather opinions, probe attitudes, gauge organizational climate, or identify improvement opportunities.

The survey is a useful technique because it is a cost-effective way to involve a large number of employees and gain useful information about the enterprise. Surveys are also useful as a means of monitoring progress in efforts to improve quality of work life or organizational climate; survey results from several successive periods can be compared as a measure of progress.

Once a survey has been administered, it is imperative that the organization do two things: feed back the results and take action on the issues identified. By administering a survey, management raises employee expectations. The work force expects to learn of the survey's conclusions and anticipates that management will take appropriate action in response to the data. If management is not prepared to provide full and honest feedback and to address the key issues raised by the survey, it would be wise to forego this technique, as the involvement benefits will largely be lost and employee resentment may well result.

A survey project at an oil refinery presents an excellent example of effective use of feedback. Following the administration of a written survey to its 2,000 employees, refinery management initially provided a written summary of results to employees. The written feedback was followed up with employee meetings, run by the refinery manager's direct reports, to provide more detailed feedback and to elicit employee reactions and ideas for dealing with the major issues. Besides receiving extensive feedback, the refinery's employees were thus involved in developing ideas for action.

Suggestion Systems

Suggestion systems are organized approaches to obtaining and evaluating employee ideas for improvements. One of the most widespread employee involvement techniques, suggestion systems unfortunately are often ineffective in practice. Ineffectiveness is not an inherent trait of suggestion systems, but instead stems from poor design. Certain features are invariably found in successful suggestion systems:

Supervisory support. Ideas are often vague and ill-defined initially, and lacking the skills to develop and articulate the ideas, employees may be reticent to submit them to the suggestion program. Supervisory assistance in defining and developing the ideas can significantly increase the quantity and quality of ideas submitted.

Prompt feedback. Lack of feedback to suggestors will ensure the failure of the system. Without disciplined feedback procedures, it will appear to employees that their ideas have disappeared into a "black hole," never to be seen again. Or worse, the idea will be implemented without appropriate credit to its creator. And feedback alone is not enough; it should be reasonably timely as well. Employees will not respond enthusiastically to a suggestion system that provides feedback eight months after submission of an idea.

Recognition and reward. Employees whose ideas provide benefit to the company are certainly not unreasonable in expecting recognition for their contributions. Recognition reinforces behaviors, and without it, the behaviors cannot be expected to recur. Many suggestion systems provide financial rewards for accepted suggestions, ranging from token amounts to substantial sums that are based on the savings realized. Successful systems typically make extensive use of nonfinancial recognition as well: suggestor's pictures appear in the company paper, for example, or monthly award dinners are held for those whose suggestions are implemented by the company.

Successful suggestion systems often incorporate innovative features to encourage and support participation. Beech Aircraft provides an example of effective use of feedback and rewards to promote a company suggestion system.[3] At Beech, all suggestions undergo an initial screening process within 3 days to determine if they are worthy of a full evaluation. If they pass this first screen, the suggestor receives an immediate $10 award. Should the suggestion ultimately be implemented following the full evaluation, a larger award, based on the savings realized, is presented. The prompt feedback and reward for simply submitting a suggestion that is not obviously ludicrous is an effective motivator; despite the existence of four full time suggestion program administrators, the system consistently returns $4 of savings for each dollar of administrative cost.

Honeywell, the manufacturer of automation and control systems, has a long history of success with a suggestion program that also employs some innovative features.[4] Established in 1942, the system has undergone a number of changes that have increased its effectiveness.

Supervisory support is ensured at Honeywell by incorporating suggestion plan activity into the performance evaluation process. Goals relating to the number of subordinate suggestions received and implemented are established for supervisors, and their success in achieving these goals is evaluated as part of the supervisor's performance appraisal.

The Honeywell program is also exceptionally rigorous with regard to feedback. The suggestor, rather than receiving a single written communication, is contacted *personally* on three different occasions: following submission, during the evaluation, and upon the suggestion's final disposition.

Finally, Honeywell employs additional incentives beyond the usual cash award based on a percentage of the first year's savings. Merchandise or gift certificates, for example, are awarded to employees as they pass certain cumulative savings milestones, as a result of multiple suggestions, during the year.

The National Association of Suggestion Systems (NASS) consistently ranks Honeywell's program as one of the most successful in the country, and this conclusion is supported by data collected by NASS. In a recent year, for example, Honeywell received 221 suggestions per 100 eligible employees, compared to an NASS average of 15 suggestions per 100 employees.

Suggestion systems can be an effective involvement mechanism, but simply hanging a suggestion box on the wall will not work. Feedback, recognition, and support are the ingredients required to ensure success.

Nominal Group Technique

The use of a structured brainstorming process like the Nominal Group Technique was described in the chapter on measurement. NGT is useful for many purposes besides measurement, however. It is a relatively quick and simple method of involving employees in addressing a variety of organizational issues, ranging from identifying improvement opportunities to establishing organizational goals.

Uncontrolled group processes suffer from a number of shortcomings. The group often loses focus, with participants discussing issues that are unrelated to the purpose of the meeting. Participation is uneven. The group often lacks an effective process for reaching consensus. And invariably, the group is dominated by one or a few individuals, either because of their rank or the force of their personalities.

The Nominal Group Technique (nominal meaning "in name only") is designed to overcome these shortcomings. The NGT approach is highly structured, and certain behaviors are not permitted. Because of these requirements, a facilitator trained in the NGT process is useful.

The group is typically selected to provide a cross section of employees, with different levels and different responsibilities represented. The diversified nature of the group ensures that a variety of viewpoints and perspectives will be represented. A group size of eight to twelve is generally most effective, although both larger and smaller groups will work.

The facilitator begins by reviewing the task statement. Sample task statements include:

☐ What are possible measures of performance for our organization?
☐ What are the major opportunities for productivity improvement in our department?
☐ What are the major barriers to employee involvement in our organization?
☐ What goals should we adopt for our group?

The facilitator may also need to provide some brief education or clarification on the subject of the exercise, particularly if the task is to develop measures. The facilitator then guides the group through the four steps of the NGT process:

1. Silent Generation. Participants are asked to ponder the task statement and to write down possible responses. This is an individual process; conferring with other participants is not permitted.
2. Round Robin. The facilitator asks each participant in turn to provide one of his ideas. Each idea is recorded on flip-chart paper as it is presented. The round robin continues until all ideas are exhausted. Participants may pass at any time, but are encouraged to re-enter the se-

quence should new ideas occur to them. The facilitator does not allow individuals to express opinions about the ideas presented; dominant participants must not be allowed to intimidate others or squelch individual creative processes.

3. Clarification Discussion. Once all the ideas have been recorded, the facilitator revisits each idea in turn, inviting participants to ask questions of clarification. The intent of this step is to ensure that all participants fully understand each item on the list. Consolidation of duplications also occurs during this step; any ideas that are different articulations of essentially the same concept are combined. Again, the facilitator does not permit the expression of individual opinions regarding the merits of the various ideas.

4. Voting and Ranking. Each participant is asked to select what he believes to be the best ideas (typically five to eight in number) and is provided with the appropriate number of index cards. The participants record their selections on the index cards, one per card, and rank them, with the highest number representing the best idea. The cards are collected, the rankings are aggregated, and the total score for each idea is posted on the flip-chart. The result is a rank-ordered list, representing the consensus of the group.

The Nominal Group Technique, by fostering individual creativity and encouraging full participation, typically results in a very long list of ideas. An NGT session conducted by the author to develop measures of performance for an engineering department, for example, resulted in 116 measures. A list such as this provides the raw material for management to develop a family of measures, as discussed in Chapter 6.

NGT is an effective and relatively simple mechanism for involving employees in a wide range of organizational issues. By tapping into the ideas of employees, management obtains a wealth of information to assist in the decision-making process. In addition, the decisions that are ultimately made are more likely to be supported by the work force since employees have provided input.

Midpoint on the Continuum

As we move down the employee involvement continuum, we encounter some additional techniques: task forces, quality circles, labor-management participation teams. These techniques are farther down the continuum because of a fundamental difference from the low-order techniques—they involve employees in problem-solving.

The techniques presented in the previous section serve either to inform employees or to gather their ideas. It is incumbent upon management to make use of the ideas or to "fix" the problems identified. The techniques in the middle section of the continuum go beyond idea generation; they engage employees in developing solutions.

Moving from idea-gathering to group problem-solving is not a trivial undertaking. Supervisors must be provided training in group leadership in order to avoid problems inherent in group dynamics. Employees must be trained in problem-solving, as they likely lack these skills. One or more facilitators are needed to organize the problem-solving groups, provide the training, and support the groups' efforts. Employees must be taken off of their jobs periodically to engage in problem-solving activities. Technical support may be necessary to help the groups deal with the technical aspects of the problems they are addressing.

Because of these requirements, problem-solving techniques mean a greater commitment of time and resources and greater management commitment to employee involvement; that is why they are farther down the involvement continuum. The extra effort and commitment required is not without a return, however. The use of problem-solving techniques can result in significant financial gains and, more importantly, can begin to foster meaningful organizational change. While the low-order involvement techniques are useful, they are unlikely to bring about a change in management style by themselves. Effective use of group problem-solving techniques, on the other hand, can lead to a more collaborative relationship between management and employees and to the institutionalization of employee involvement in problem solving on the job.

However, because problem-solving techniques are more sophisticated and require greater support than the idea-gathering techniques, the organization must consider its state of readiness. If the company has not had good results with the low-order techniques and has a low awareness of the nature of employee involvement, it would probably be wise to defer its use of problem-solving techniques until experience has been gained and the organization has been prepared through the use of simpler and less demanding techniques. More than a few quality circle programs have failed to meet expectations and have ultimately failed because the organization was not ready for that degree of change.

Task Forces

The least demanding of the problem-solving techniques is the task force. A task force is simply an *ad hoc* group of managers and employees that have been brought together to solve a particular problem. The problem is defined by management and the group is typically disbanded after a solution has been developed and presented. Some group leadership and problem-solving training is often provided, but not always.

Since a task force has only a limited life and little or no discretion in choosing its problems, it is a relatively low-risk means of engaging employees in problem-solving. As such, it can serve as a useful bridge between the low-order involvement techniques and the more sophisticated problem-solving techniques. Through a succession of task forces, management can gain experience in the use of problem-solving techniques and prepare the organization for permanent problem-solving teams.

One company that has made extensive use of task forces is Champion International.[5] Individual paper mills have involvement team coordinators, who organize employee teams to deal with specific, identified problems. Team members, usually eight to ten in number, are selected by supervisors. The team is provided with a half-day of training, primarily covering the nature and benefits of involvement. Solutions developed during regular meetings are presented to management for approval. In a variation from the usual task force approach, Champion does not automatically disband a team once its problem is solved, but may provide it with an additional problem to solve. Successful teams thus may have an indefinite life. Annual savings in excess of $1 million have been reported by some Champion mills through the use of task forces.

Quality Circles

Quality circles are permanent problem-solving teams consisting of employees from a common work area. Participation in quality circles is voluntary, and the circles meet on a regular basis on company time to identify and solve problems. Circles choose their own problems to solve and thus enjoy a greater degree of autonomy than the task force. Despite the name of this technique, problems addressed are not limited to quality, but typically include productivity, quality of work life, and other organizational performance issues as well.

Generally, the supervisor of the involved work area is the circle leader. Since the typical supervisor is not well-versed on group dynamics, it is vital that he receive group leadership training. The supervisor must know how to enlist full participation, provide feedback, handle conflicts, and manage problem-solving activities.

Circle members also require training, particularly in problem-solving skills. The average nonmanagement employee does not know how to gather and analyze data and to develop solutions to problems. Techniques typically taught in problem-solving training include Pareto Analysis (the application of the 80/20 rule), Cause-and-Effect Analysis, and Statistical Analysis. Employees may also need to develop presentation skills, as the circle's proposed solutions must be presented to management in a convincing fashion.

Quality circle programs invariably require one or more facilitators to manage them. The facilitator recruits volunteers, starts up circles, provides training, handles logistical matters, ensures the availability of technical support, coaches circle leaders, and, in general, manages the effort. The facilitator, of course, must be trained in program management and facilitation as well as receive all of the training provided to circle leaders and members.

Given the significant training needs and the added personnel requirements (for facilitators), it should be clear that a quality circle effort is a major undertaking, certainly exceeding the resource and management commitment requirements of the low-order involvement techniques.

Nonetheless, quality circles enjoy enormous popularity in American industry and seem to be ubiquitous; companies routinely brag of having quality circles numbering in the hundreds, and even thousands. And therein lies a problem: quality circles, more than any other technique, epitomize the Techniques Trap, the unfortunate tendency to equate employee involvement (or even productivity improvement) with a given technique, losing sight of the broader organizational change objectives.

Not that the quality circle is not useful; on the contrary, it can be a very effective employee involvement technique. But it is *only* a technique, and its value lies in its use as a mechanism to foster change. It does not, by itself, constitute employee involvement, and the organization that seeks simply to maximize the number of quality circles implemented has missed the point.

Other Problem-Solving Techniques

The basic quality circle model has been modified in an almost infinite variety of ways by American companies. Some of the more popular spin-offs include:

- ☐ Labor-management participation teams. A staple of labor-management cooperation efforts (Chapter 13), LMPTs are quality circles whose membership includes both union and management employees.
- ☐ Cross-functional teams. With various departments or functions represented, these problem-solving groups are designed to deal with organizational interface problems.
- ☐ Business teams. Organized around product lines or markets, these groups deal with marketplace or product-specific issues.
- ☐ Horizon teams. Developed by Xerox, these groups address long-term planning issues.

High-Order Involvement Techniques

Many organizations now sit firmly at the mid-point on the involvement continuum, making extensive use of group-based problem-solving techniques. But since these approaches are only techniques, they cannot be relied upon to last forever. In fact, many companies with quality circle programs find that the organization's appetite for this technique is satiated after three or four years, and enthusiasm wanes. Like any technique, they do not last forever.

The organization that has not considered its next evolutionary phase of development has a problem if quality circles fade away. It no longer has an effective employee involvement technique, other than low-order ones, and has thus taken a step backwards. Hopefully, the effort has yielded some very positive results—an improved climate and improved performance—but where does it go from here if it wishes to further develop the desired culture of teamwork and employee participation in performance improvement?

There are indeed more sophisticated techniques for involvement beyond quality circles. To reach the far end of the continuum—and the ultimate in employee involvement—we must incorporate the principles of employee involvement into day-to-day operations and, ultimately, into the structure of the organization and the very design of work itself.

The problem-solving techniques occupying the mid-point on the continuum are not actually integrated into the structure and practices of the organization. Involvement takes place "off-line," rather than within the context of employees' day-to-day jobs. The problem-solving groups are not found on organization charts, but are actually *parallel* to the formal organization (hence the name parallel structures).

If we wish to achieve the ultimate in employee involvement, commitment, and effectiveness, we must redesign jobs, and the organization itself, so that maximum involvement is built into day-to-day work practices. As an interim step, we can achieve this integration within the traditional organization and job design (see "natural work teams" below), but if we wish to truly maximize involvement, we must redesign jobs and the organizational structure itself to accommodate maximum involvement.

The highest-order techniques, then, may be characterized as *work redesign*. This represents the far end of the continuum, the state-of-the-art in employee involvement. The commitment to employee involvement must be extremely high, because these techniques require radical change from traditional and comfortable ideas about management. They are not for the novice and should not even be considered by organizations that are just learning what employee involvement is all about.

While the requirements and the risk are great, the return is commensurate. Organizations that have successfully implemented work redesign efforts have realized enormous, and sometimes startling, improvements over the performance of similar facilities that are traditionally designed and managed.

For those organizations not ready or willing to undertake the radical change implied in work redesign, the natural work team approach may be more appropriate.

Natural Work Teams

Incorporating involvement activities into the existing organizational structure and processes is a natural progression from the use of parallel structures. While technically not work redesign, since jobs and the organizational structure may not be radically modified, the natural work team approach nonetheless begins to institutionalize employee involvement by incorporating involvement practices into the day-to-day activities of the organization.

Implementation of this approach starts with identification of the natural work groups within the organization. A natural work group consists of a number of employees who are organized around a single work process, are interdependent in their functioning, and have a common supervisor. Everyone in the organization is a member of a team, and teams extend all the way up the hierarchy. A supervisor or manager is therefore both a team leader and a team member, as graphically represented in Figure 8-2. All employees are trained in problem solving, and all supervisors and managers are trained in group dynamics and team leadership.

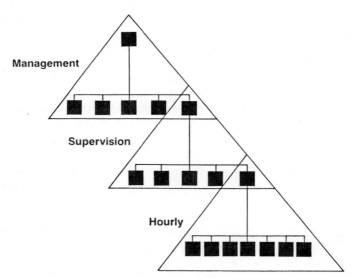

Figure 8-2. Natural work teams.

Natural work teams represent true participative management. Supervisors conduct regular communications meetings in order to keep their subordinates fully informed regarding business conditions, organizational performance, and competitive challenges. Employees are engaged in problem solving as problems arise. Supervisors solicit employee ideas for performance improvement and informally recognize employee contributions. There is a sense of common objectives, and employee commitment is high.

The natural work team approach cannot succeed without a true commitment to employee involvement on the part of supervisors and managers. Since participation is not highly structured, as it is in the low-order techniques and the parallel structures, it will occur and be sustained only if the supervisor believes it is the best way to manage his subordinates, has the necessary skills, and enjoys the support of higher levels of management. Top management can mandate quality circles (although they probably won't last), but they cannot mandate participative management on the job.

Since participative management of natural work teams represents the ultimate goal of many employee involvement efforts, why should organizations bother with the intermediate step of parallel structures? The answer is that parallel structures such as quality circles enable the organization to learn and practice participative management skills "off-line," without disrupting day-to-day business. Supervisors gradually become acclimated to a different role and employees gain proficiency in problem solving. When the transition is made from parallel structures to natural work teams, the organization and its employees are ready.

Autonomous Work Teams

The best known approach to true work redesign is the autonomous work team. The autonomous work team is organized around a natural work process, but differs substantially from a traditional work group. In traditionally designed organizations, we divide the work associated with a given process into various tasks and assign each task to a particular employee. Some employees are machine operators, for example, while others are maintenance people, quality control technicians, or schedulers. In the autonomous work team, by contrast, employees are cross-trained to execute all of the jobs associated with a given process.

The reward system supports this job design. Rather than a traditional system that provides for increases based on seniority, we typically find instead a pay-for-skills (or pay-for-knowledge) system. An employee's advancement into higher job classifications—and associated higher pay levels—is dependent upon his becoming proficient in the various skills associated with the work process. In other words, as the employee learns each new skill, he advances into a higher pay classification. High involvement systems also commonly make use of group incentive systems, such as gain sharing (Chapter 10), to more closely tie compensation to organizational performance.

Perhaps the most noteworthy departure from traditional job design is hinted at in the name of the technique: *autonomous* work teams. These are essentially self-managed work units. Members of an autonomous work team are involved, on a routine basis, in decision-making, goal-setting, scheduling, hiring, planning, peer reviews, and problem-solving.

They determine how they will work together, for instance, and how they will rotate through the various jobs associated with the process. They have a hand in developing performance measures for their group, and in setting goals around these measures. They are intimately involved in the hiring process; new members cannot be brought into their team without their approval. They manage conflicts within the group and deal with disciplinary problems. They constantly monitor their progress and solve problems as they arise.

The selection process is typically very rigorous in these organizations. For example, management of the Rohm and Haas chemical plant in LaPorte, Texas identified certain qualities that they felt were the personal attributes required to support their high-involvement system.[6] Definitions of these attributes were incorporated into a structured interview process involving all workers in the area where a job opening exists. Their list included:

☐ **Responsibility.** Maintaining a high standard of work performance, dedication to excellence/quality, setting goals for self and organization.
☐ **Motivational Match.** Congruence of the person's needs, preferences, and satisfactions with the demands, opportunities, and requirements of the job.
☐ **Versatility.** Handling multiple tasks, interruptions, and diversity in work setting.
☐ **Learning Ability.** Assimilating and applying new job-related information.
☐ **Honesty.** Maintains social, ethical, organizational norms in job-related activities.
☐ **Self-starting.** Active efforts to influence events and keep oneself involved in productive activity.
☐ **Cooperation.** Working as a team member to ensure that team goals are accomplished; deference to team goals over individual activities.
☐ **Openness.** Genuineness in dealing with others; willingness to relate openly to others and meet interpersonal conflict as a mature adult.
☐ **Tact/Sensitivity.** Aware of, responsive to the impact of his behavior on the feelings, dignity, and performance of others.

Supervisors still exist in these organizations, but their role is very different from that of the traditional supervisor. The supervisor's job is not to direct and control his subordinates, but to develop his team's self-management capabilities. The supervisor's job is not done until his team can effectively function without him. The supervisor also has the ongoing responsibility of managing the interfaces between the teams, as the organization will likely consist of a series of interlocking teams.

The autonomous work team is capable of very high levels of performance because it maximizes employees' capabilities to contribute to organizational performance and fosters an extremely high level of employee commitment. Employees are no longer treated as cogs in a machine, taking orders from a domineering supervisor, but are accorded the opportunity to manage themselves. In effect, they are proprietors of their own little business within the larger organization.

The success of this type of organization is also attributable to the fact that it is extremely lean. Because of the self-managing design, supervisors can effectively handle very large spans of control, and therefore very few management layers are necessary. It is not unusual for a large plant to have a single layer of management between the plant manager and the hourly worker.

HIGH-INVOLVEMENT SYSTEMS IN PRACTICE

Experimentation with the autonomous work team and related approaches has been underway in the United States for several years. One of the first experiments was in the early 1970s at a General Foods plant in Topeka, Kansas. Companies such as Procter and Gamble and Cummins Engine are well-known for their use of these techniques in new plant start-ups. General Motors, in their well-publicized effort to create a small car manufacturing subsidiary with state-of-the-art technology and organizational design (dubbed the "Saturn" project), concluded after much study that the autonomous work team concept was the most effective approach possible to job and organization design. Characterized by a major business periodical as "the boldest experiment ever in self-management and consensus decision-making," Saturn is designed to operate without foremen and to employ consensus decision-making between management and the union.[7] While the project was subsequently deemed too ambitious technologically and was accordingly scaled back in 1986, the innovative work design features remained largely intact.

All of the features of the autonomous work team concept can be seen in a Sherwin-Williams paint plant in Richmond, Kentucky.[8] Born of the need to respond to increased competitive challenges, the Richmond plant was an attempt to consolidate several existing operations into a modern facility with a work environment designed to maximize the contributions of its workers.

The philosophies underlying the organizational design were articulated in the Richmond Charter (Figure 8-3).

As in the Rohm and Haas example, the selection process at the Richmond plant is exhaustive. All job candidates are carefully screened to ensure that character traits are consistent with the ideals espoused in the Richmond

The Richmond Charter

Purpose: To construct and operate a safe, clean, efficient plant which will produce the highest quality automotive refinish paint in the world and keep itself ahead of the industry in competitiveness and profitability.

People: We expect to employ mature, responsible and cooperative people . . . who want to work in an open and trusting climate and who want to participate as responsible employee/business partners on a continuing basis to improve themselves and the effectiveness of the plant.

Jobs: We will develop a safe, clean and healthful working climate that provides challenging and meaningful work with the opportunity for personal growth and development.

Compensation: We will have a fair and equitable compensation system that rewards all plant personnel on the basis of job knowledge, performance against goals/objectives and training skills.

Plant Management: We operate . . . in a manner that demonstrates good communications, respect for people, honesty, openness and a responsiveness to realistic ideals and suggestions from both the plant personnel and the community. We are expected to meet the objectives for both short and long range goals, which will be communicated to all employees.

Sherwin-Williams Co.: We expect the plant to be profitable, contributing to Sherwin-Williams' financial growth.

Figure 8-3. The Richmond Charter.

Charter. Applicants are interviewed not only by management, but also by a committee of workers in the team with the job opening.

Every employee at the Richmond plant is a member of one of ten teams (Figure 8-4). Each team sets goals on a variety of performance measures tailored to its work area. Team members are involved in decision making on such issues as shift rotation, budget development, vendor relations, peer appraisals, equipment selection, work assignments, inventory levels, preventative maintenance, equipment repairs, and problem resolution.

As with most autonomous work team installations, the Richmond plant employs a "pay-for-knowledge" reward system. A new employee enters at

Customer Service: We expect to gain a reputation as a dependable, fair and honest supplier of automotive refinish paints by developing within the plant a working knowledge of our customers' needs and their use of our products. We encourage communications between plant personnel and these customers.

Training: We see training and development as a day-to-day activity in which each plant employee participates both as a learner and as a teacher.

Personnel: We will have fair and understandable personnel policies which work toward filling the needs of employees, the plant and the community, as well as providing guidance and assistance where needed.

Purchasing and Supplies: We expect to deal with reliable suppliers who provide quality material and services at the least cost that is consistent with the quality and service provided. We expect to build strong relationships with our suppliers that will encourage their serving as part of an effective plant team.

Family: We want our families to understand the plant and its importance to them, the community and Sherwin-Williams; to feel pride in being associated with the company; and to feel they are better off as a result of its work here. We want the economic needs of our families to be met by providing safe, clean, stable employment that also provides good financial compensation and benefits.

Community: We want the Sherwin-Williams Co. to be recognized within the community as an active, solid citizen and a good neighbor.

Figure 8-3. Continued.

an entry-level salary (all employees are salaried) and receives an increase when he has mastered his first job. Subsequent pay increases are awarded as the employee masters additional jobs in his department's work area. Both written and proficiency tests must be passed in order to qualify for the higher salary. Once all jobs are mastered, the employee begins rotating through all of the team positions.

The Richmond plant also employs a gain sharing, or group incentive, system (Chapter 9). Bonuses are paid to employees during any month in which plant costs are more than one-fourth of one percent below budgeted cost levels. The company retains 25% of the gain with the balance going to employ-

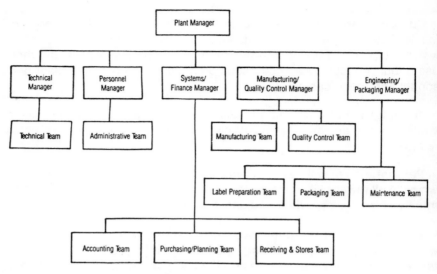

Figure 8-4. Sherwin-Williams Richmond Plant organization.

ees. A portion of the employee share is held for a time in a reserve or contingency fund to help offset the negative financial performance associated with months when costs exceed budget. The design features of gain sharing systems are discussed in detail in Chapter 10.

An incident that occurred in 1984 should dispel any doubt about whether high-involvement systems engender employee commitment and loyalty. The Richmond plant workers surprised management with a partial give-back of their first-half productivity bonuses for that year, citing their appreciation for the company's providing them with continuous employment and excellent pay and benefits during difficult economic times.

Use of the autonomous work team concept is not limited to manufacturing. Two pilot work teams were established in the collections department of Continental Illinois National Bank's credit card operations.[9]

Each pilot team consisted of ten collectors and an advisor, whose role was to be a trainer and a resource to the group rather than a boss. The teams were self-managed, and the experiment included such features as peer appraisal and team determination of salary increase distribution.

Results were mixed, with one pilot producing moderately positive results and one failing. The one that succeeded did so even in the face of a major distraction—the sale of Continental's credit card business to Chemical New York Corporation. The second pilot's failure was attributed in part to inadequate training, which left team members and advisors inadequately prepared for a major change in orientation.

Another example of high-involvement job design can be found in the federal government. The U.S. Copyright Office began an employee involvement process in 1980 that included experimentation in redesigning jobs along sociotechnical lines.[10] The initial pilot effort was in the area where copyright title transfers were processed. Employees were organized into semiautonomous work teams with responsibility for handling the entire process, including receipt of applications, examination of works for copyrightability, handling of correspondence, and registration of items. Prior to the redesign, records might pass through as many as ten work units before processing was complete. A significant result of this initiative was the reduction in turnaround time from as long as four months to about two weeks.

Despite the fact that work innovations of this nature are hard to find in a service business, there is no compelling reason why the autonomous work team concept cannot succeed in a white-collar environment. If one believes that people are the key to success, and that maximum organizational performance is a function of the degree to which employees are allowed to utilize their capabilities and to exercise discretion in the conduct of their jobs, those principles should have validity regardless of the nature of the work process.

USING MULTIPLE TECHNIQUES

One organization that has effectively utilized a variety of techniques is the Aerospace Products Division of Lord Corporation, a privately-held manufacturer headquartered in Erie, Pennsylvania.[11] In 1983, the company initiated a Quality/Productivity Improvement (QPI) effort with a focus on employee involvement.

Recognizing that the attitudes, skill levels, and readiness for involvement varied from one part of the organization to another, division management structured five different types of "involvement groups" (Figure 8-5). The groups ranged from one that is beyond the parallel structures point on the continuum (Type 1) to one that is basically limited to communications meetings (Type 4) and thus represents a low-order form of involvement. There is also an ad hoc group (Type 5), that is basically a task force.

Each work area is evaluated based on several criteria to determine which technique is most appropriate. Provisions are also made for evolution to higher-order groups when the work group displays the appropriate readiness.

IMPLICATIONS OF THE CONTINUUM

The concept of an employee involvement continuum provides two lessons for the organization seeking to improve productivity through people:

Type 1

Objective:
- Full implementation of QPI philosophy/tools and employee involvement

Characteristics:
- Machine/operation grouping
- Labor classification combination
- Selected supervisor/operators
- Cross-shift participation
- Broad operator independence
- Broad problem scope
- Daily involvement
- High organizational visibility
- Broad interaction with Lord/Customer/Vendor personnel
- Ongoing training

Type 2

Objective:
- Implementation of QPI philosophy/tools and inter-functional employee involvement

Characteristics:
- No initial change to machine/operations
- No classification combination, but more than one involved
- Incumbent supervisor/operators used
- May be cross-shift participation
- Some operator independence
- Somewhat broad problem scope
- Generally meet bi-weekly
- Some organizational visibility
- Some interactions with Lord/Customer/Vendor personnel
- Periodic training

Type 3

Objective:
- Implementation of QPI philosophy, some use of tools and intra-functional employee involvement

Characteristics:
- No initial change to machine/operations
- Generally one classification involved

Figure 8-5. Types of hourly involvement groups, Lord Corporation.

- Incumbent supervisor/operators used
- Generally single shift participation
- Some operator independence
- Focused problem scope
- Generally meet bi-weekly
- Low organizational visibility
- Occasional interactions with Lord/Customer/Vendor personnel
- Periodic training

Type 4

Objective:
- Communication of QPI philosophy and limited use of tools

Characteristics:
- No change to machines/operations
- More than one classification may attend
- Incumbent supervisor/operators attend
- Generally single shift participation
- Little operator independence
- Mostly informational/little problem involvement
- Generally meet monthly/bi-weekly
- Rare organizational visibility
- Rare interactions with Lord/Customer/Vendor personnel
- Occasional training

Type 5 (Ad Hoc)

Objective:
- Implementation of QPI philosophy/tools and both inter- and intra-functional employee involvement to resolve individual problems

Characteristics:
- No initial change to machines/operations
- More than one classification may participate
- Selected supervisor/operators with knowledge to correct problem
- Generally single shift participation
- Little operator independence
- Focused single problem scope
- Meet as required to resolve problem
- Generally low organizational visibility
- Occasional interactions with Lord/Customer/Vendor personnel
- Training required to resolve problem

Figure 8-5. Continued.

1. The "best" technique for involving employees is situational and dependent upon the organization's present climate, level of management commitment, and past experience with employee involvement.
2. Employee involvement is a constantly evolving and growing process, with ultimate success dependent upon the company's ability to consciously increase the level and sophistication of its involvement activities.

In short, the organization seeking to develop a productivity management process must intelligently select the techniques it utilizes to involve its employees and must move down the involvement continuum in a systematic and rational fashion. Forcing the use of a technique for which the organization is not ready is a recipe for failure. At the same time, the reliance on a single technique, even if it is "right" at the time, will not provide for the organization's necessary and continuing development toward a more effective management style.

It is not difficult to find examples of companies whose employee involvement efforts have lost momentum after a period of years because they became "stuck" at some point on the continuum. This problem is often encountered by companies caught up in the quality circle movement of the 1980s and, as was suggested earlier, is symptomatic of the Techniques Trap (Chapter 5).

Techniques come and go, and the organization's appetite for any particular technique is ultimately satiated. Any effort to transform management style and culture requires flexibility, constant revitalization, and deliberate movement toward greater levels of involvement and the institutionalization of involvement in the day-to-day work practices of the organization.

Movement down the continuum does not have to occur in great leaps, but can be an evolutionary process. For example, once the organization has become proficient in involving employees in problem-solving, through quality circles or some other technique, it may then begin to involve employees in goal-setting. From there it may seek employee contributions to scheduling, hiring, or planning decisions. It is moving down the employee involvement continuum a step at a time.

REFERENCES

1. Excerpted from the Computer Conference on Reward Systems, sponsored by the American Productivity Center as part of the White House Conference on Productivity, 1983.
2. *Case Study 49: Allen-Bradley*. Houston, TX: American Productivity Center, 1985.

3. *Case Study 6: Beech Aircraft Corporation.* Houston, TX: American Productivity Center, 1980.
4. *Case Study 13: Honeywell, Incorporated.* Houston, TX: American Productivity Center, 1980.
5. *Case Study 18: Champion International.* Houston, TX: American Productivity Center, 1980.
6. *Case Study 51: Rohm and Haas Bayport, Inc.* Houston, TX: American Productivity Center, 1986.
7. "How Power Will be Balanced on Saturn's Shop Floor." *Business Week*, August 5, 1985.
8. *Case Study 40: The Sherwin-Williams Co.* Houston, TX: American Productivity Center, 1984.
9. "Chemical Financial Services Corp's Attempt at Autonomous Work Teams Yields Few Successes, Several Lessons." *Productivity Letter.* Houston, TX: American Productivity Center, June 1985.
10. *Case Study 41: U.S. Copyright Office.* Houston, TX: American Productivity Center, 1984.
11. *Case Study 55: Lord Corporation Aerospace Products Division.* Houston, TX: American Productivity Center, 1986.
12. Eitington, J. E., *The Winning Trainer: Winning Ways to Involve People in Learning.* Houston, TX: Gulf Publishing Company, 1984.

9

Reinforcing Reward Systems

REWARDS AND PRODUCTIVITY

Reward systems that explicitly and consistently reinforce productivity improvement are essential to the productivity management process. Reward systems are a major determinant of individual and group behavior, as people usually behave in ways that they believe will bring them greater rewards or enable them to avoid punishment. As a result, the impact of reward systems on organizational performance is enormous.

If we wish to create an organizational culture in which productivity is a driving force and a way of life, reward systems must play an important role. How can we expect people to behave in a manner that is consistent with our productivity management process if we don't reward the desired behaviors?

Unfortunately, the reward systems of most traditionally-managed organizations fail to effectively reinforce productivity improvement. One of the biggest culprits is the merit increase system. Designed to provide employees (at least salaried ones) with financial rewards for personal performance, most merit systems simply fail to achieve that objective. Not that the concept is flawed; the problem generally lies in the execution. The company typically budgets for a given percentage increase in salaries (usually approximating the anticipated inflation rate for the upcoming year), and managers distribute their portions of the merit increase budget among their employees. Since the money available for increases is predetermined and fixed, any greater-than-average increase must be offset by a comparably lesser-than-average increase to another employee reporting to the same manager. Awarding a subordinate a small (or worse yet, nonexistent) increase is distasteful to most managers, and since there are generally few objective performance measures available to justify such a decision, the typical manager avoids the problem by clustering the increases tightly around the average.

This situation is often exacerbated by policy restrictions and extraordinary justification requirements associated with the granting of unusually high increases. The result is that the merit system essentially serves as a cumbersome mechanism to distribute annual cost-of-living increases to the company's employees. Is it any wonder that employees scoff when a company utilizing this approach maintains that it has a "pay-for-performance" system?

Another organizational system that rarely realizes its full potential as a reinforcing mechanism is the performance appraisal system. An interesting and revealing activity—one that provides a quick gauge of the status of productivity in an organization—is to review the company's standard performance review form. Specifically, the observer should seek to find the word "productivity" somewhere on the form. The odds are high that the word will not be there.

The nonreinforcing nature of our traditional reward systems is even more glaringly apparent with respect to the hourly-paid work force. While systems usually exist that at least have the potential to reinforce productivity improvement by salaried employees, often no such mechanism can be found for the hourly employee. Wage increases are typically general or contractual in nature, and performance reviews for hourly employees are the exception rather than the norm.

The problem extends well beyond the realm of financial rewards. Most organizations also do a poor job of reinforcing performance through nonfinancial rewards, such as recognition. The author long ago lost count of the number of times he has heard the phrase, "The only time you get any recognition around here is when you screw up!" Informal, on-the-job recognition is simply not a behavioral norm of the traditional supervisor. Employees are paid to do what they're told, and the supervisor has more important things to do than run around patting his subordinates on the back all day.

Reward systems, both financial and nonfinancial, are in a sorry state in most organizations. They are wholly inadequate to support any attempt to develop a productivity management process. They not only fail to reinforce productivity improvement, but they may actually reinforce nonproductive behaviors.

If productivity is to advance beyond the "program" stage, we must modify our reward systems to explicitly, continually, and effectively reinforce productivity improvement.

REWARD SYSTEMS CLASSIFIED

Organizations generally fail to capitalize on the full range of reward systems available to reinforce productivity improvement. Managers tend to perceive the subject very narrowly, thinking only in terms of individual-oriented financial rewards. To ensure consideration of the full range of options, it is useful to categorize reward systems in different ways.

One way to categorize reward systems is *financial versus nonfinancial.* Financial rewards are those that provide direct cash remuneration to employees. Salaries, wages, commissions, piecework systems, executive bonuses, profit sharing, and merit increase systems are well-known examples of financial reward systems. Others that may not occur to the manager as readily are well pay (additional compensation provided for excellent attendance) and suggestion system awards (a percentage of the savings realized through submission of individual improvement suggestions). There exists within the area of financial rewards a number of innovative practices that are growing in popularity: lump sum increases, pay-for-knowledge systems, all-salaried work forces, and gain sharing. These approaches will be discussed later in this chapter.

Nonfinancial rewards also take a variety of forms. All of the following can be viewed as nonfinancial rewards:

☐ **Recognition.** The best known and most prevalent form of nonfinancial reward, recognition practices range from formal awards accompanied by much hoopla to simple pats-on-the-back by supervisors.

☐ **Responsibility.** Enlarging an employee's job responsibilities or freeing him from close supervision provides reinforcement for regular and continuing high performance.

☐ **Growth.** Providing employees with opportunities to learn new skills and to practice those skills not only improves employee capabilities, but also serves as a reward.

We might also categorize reward systems as *individual versus group* rewards. The reward systems in most American organizations are heavily weighted toward the reinforcement of individual initiative and performance. With the exception of profit-sharing, all of the most common forms of financial reward (see previous section) are individual-based. This phenomenon effectively skews behavior toward the maximization of individual performance, often to the detriment of group results. This is not to suggest that individual performance should not be rewarded; rather, what is needed is more balance between individual and group reinforcement. The rapid rise in the popularity of gain sharing (a system that pays employees bonuses based on organizational performance) is due in part to a recognition of the need to create more collaboration and cooperation within organizations.

Nonfinancial rewards can also be either individual or group based, but here again, we see a heavy skewing toward the individual end. Recognition of group accomplishments can, like group-based financial rewards, serve to foster a more collaborative environment within the organization.

INNOVATIONS IN REWARD SYSTEMS

As more and more companies have sought to create an organizational focus on productivity, recognition of the failings of our traditional reward systems has grown commensurately. The result has been heightened interest and experimentation with reward systems that incorporate new features and depart from the traditional mode.

The American Productivity Center, in cooperation with the American Compensation Association and some corporate sponsors, conducted a major survey in 1986, covering about 10 percent of the entire civilian working population of the United States, on the use of innovative reward practices.[1] Among the major findings were the following:

☐ There has been a striking growth in the number of firms adopting nontraditional reward systems during the last five years. For example, almost 73 percent of existing gain sharing systems have been implemented since 1980.
☐ The more competition a firm reports, the more likely it is to use a nontraditional reward system.
☐ All of the nontraditional reward plans are reported to have a positive impact on performance.
☐ The number of firms not using but planning to use nontraditional reward systems indicates a continuing trend toward increased usage.

Some of the innovative approaches improve quality of work life, while others directly reinforce performance or productivity improvement. Several are discussed below.

The All-Salaried Work Force. Managers in many organizations have come to view the line between salaried and hourly workers as artificial, demoralizing, and destructive. The distinction clearly creates two classes of workers: one that is presumed to be reliable and trustworthy, and another that is presumed to be irresponsible and devious. Salaried employees don't punch a time clock because they are solid citizens that can be trusted to be at work on time. Hourly employees, on the other hand, must be monitored carefully or they will surely take advantage of the company's laxness.

The division of employees into two classes seems inconsistent and incompatible with the intent of the productivity management process—to create an environment within which all employees are working together toward improved performance. The two-class system is clearly incompatible with such an environment.

The obvious solution is to create a single class of employees—the all-salaried work force. By so doing, management sends an unmistakable message:

that it considers all of its employees to be mature adults, capable of behaving in a responsible manner. Management, in short, trusts its employees. The environment now is very different; an assumption of irresponsibility has been replaced by an assumption of trust.

As with any change, there are potential trade-offs that must be considered. There is the possibility that compensation costs will increase, as the organization no longer docks employees for tardiness or absences. An interesting phenomenon often occurs, however: Tardiness and absenteeism decline, reflecting employees' heightened commitment to the organization.

The company utilizing this approach must also continue to manage attendance, as there will always be a certain number of employees who will violate the trust accorded them. It would be argued by many, however, that the proportion of employees demonstrating such irresponsible behavior is no greater than it is within the traditional salaried workforce.

Cafeteria Benefits. Fringe benefits in the vast majority of organizations are predetermined and fixed. All employees receive the same package of benefits, whether they need them or not.

Unfortunately, needs vary greatly from individual to individual. The young, single employee, for example, probably cares little about retirement benefits but may relish free time away from work. The employee with a large family, on the other hand, may put a premium on medical and life insurance. And the elderly employee may wish to capitalize on savings programs that contribute to a more secure retirement.

Cafeteria benefits are designed to meet the varying needs of different employees. Rather than providing a fixed benefit package, the organization commits to provide its employees with benefits up to a maximum level of cost. Employees then select from a menu of benefits, with their associated costs, the combination that best meets their needs. One employee may select a high level of life insurance, while another employee may opt for more vacation time. The only limitation is that the total cost of the benefits selected must not exceed the stipulated maximum.

The primary benefit of this approach, of course, is that it enables employees to tailor the benefits package to their individual needs, thus improving their quality of work life.

The primary disadvantage lies in the administrative costs associated with this program; systems must be developed to administer this rather complex approach. It also is not practical for smaller organizations, as some benefits, such as health insurance, become considerably more expensive if only a small number of employees participate.

Pay-for-Skills Systems. In contrast to traditional hourly worker reward systems, which tie wage increases to seniority (the employee moves into higher pay classifications because of his senior position relative to other applicants for the position), pay-for-skills systems relate compensation to the acquisition of skills. Also called pay-for-knowledge, these systems reward employees for enhancing their value to the company by broadening their repertoire of skills.

Pay-for-skills systems are normally associated with the autonomous work team concept (Chapter 8). As was discussed earlier, this approach to organizational design requires that employees be cross-trained to perform the various jobs associated with a given work process. A pay-for-skills system is ideally suited to this type of work design, as it encourages and reinforces the acquisition of the necessary skills to effectively execute the various tasks required of the team. Both the Sherwin-Williams and the Rohm and Haas plants described in the previous chapter utilize this form of reward.

The employee typically starts with the company at an entry-level pay classification and begins to learn and apply his first job skill. When proficiency in that skill has been demonstrated, either through a test of some kind or to the satisfaction of his peers, the employee advances to the next higher pay classification. He may then set about learning an additional skill, which, when acquired, advances his pay again. The cycle continues until the employee has all of the necessary skills and has reached the highest pay classification for his area. Future pay increments beyond that point may come through general increases, merit increases, or gain sharing (see next section).

Because pay-for-skills systems are normally associated with high involvement, team-oriented organizations, the requirements for advancing to higher pay classifications typically include more than just job proficiency; there may also be requirements relating to the employee's ability to function as part of a team and to contribute to broader organizational objectives. These requirements are apparent in the compensation system guidelines developed by the Gaines dog food plant in Topeka, Kansas. The Gaines plant represents one of the earliest and longest-running of the autonomous work team installations in the United States. The Topeka plant guidelines are presented here:

TOPEKA PLANT
COMPENSATION SYSTEM GUIDELINES

Rate progression is dependent upon the acquisition and development of a broad range of technical and interpersonal skills. These skills are necessary for the continuity of both the plant's operating performance and the Topeka organizational concept. Skill areas include maintenance, quality systems, safety, housekeeping, plant operation, training, and team contribution.

Learning and Operation:

Rate achievement will be satisfied when an individual has successfully fulfilled the following criteria to the satisfaction of the team and the team leaders.

Operating—A person's knowledge of their area:

A. Ability to operate systems efficiently
B. Knowledge of product technology
C. Ability to troubleshoot and correct operational problems
D. Suggests improvements and follows through with them
E. Maintains open communication with other areas and teams

Maintenance—How does a person maintain their equipment:

A. Maintenance ability
B. Preventive maintenance
C. Corrective maintenance

Training—A person's performance as a trainer:

A. Willingness to share their knowledge, experience, and ideas
B. Ability as a trainer

Team Contribution—What does a person contribute to their team:

A. Does the person strive to improve themselves and their team, and does this effort result in a valuable contribution?
B. Is this person there when needed and helpful when present?
C. Demonstrates a sense of patience, understanding, and compassion for his fellow teammates
D. Applies peer pressure in an attempt to improve proficiency of others
E. Willingness to give time
F. Recognizes and responds to problems in other areas
G. Effectively participates in evaluations

Quality Awareness—Does a person take the necessary steps to ensure a quality product?

A. Does a person keep their area clean, orderly, and sanitary during production and on weekend clean up?

Downtime—How does a person utilize their time when:

A. His line goes down
B. In a nonproduction situation
C. Doing weekend work

Safety—Is the person safety conscious?

A. Does the person follow plant safety rules?
B. Does the person inform others of their infractions of those rules?
C. Does the person use good judgment and show little carelessness?

Other Team Input—How do the other teams view the individual and his job performance?

Team Meetings and Committee Work—How does a person perform on a committee or in a team meeting?

A. Does the person make a valuable contribution to the committee or team?
B. Does the person adequately represent his team's input to the committee?

A Maxwell House plant in Houston has applied the pay-for-skills approach to its first-line supervisors as part of an effort to redesign its organization to the autonomous work team concept. Supervisors progress through four salary grades as they gain and demonstrate certain proficiencies:

1. Small group leadership
2. Autonomous work team facilitation
3. Management of new department start-ups
4. Business management of a manufacturing department

While pay-for-skills systems are often thought to be associated primarily with manufacturing organizations, a recent study[2] provided data to the contrary. A national probability sample of 154 corporations revealed that 12 have pay-for-skills systems in at least one location; of those 12, only 5 were manufacturing organizations. The rest included companies in the transportation, communications, finance and insurance, and wholesale and retail trade industries.

The principal advantage of a pay-for-skills system is that it increases the capabilities of its employees and provides the organization with greater flexibility in deploying its resources. The approach can also contribute to a higher quality of work life, as employees have the opportunity to develop and grow, and they enjoy greater job satisfaction by performing a variety of tasks. A disadvantage of this type of reward system is that it requires considerable investment in training support. If the organization wishes its employees to acquire a variety of skills, it must be prepared to provide the necessary training and resources to facilitate that process. The pay-for-skills approach may also result in higher overall compensation costs, as employees will generally advance to the highest pay classification more quickly than would occur under more traditional reward systems. The benefits of the higher skill levels associated with this system, however, should more than justify the added cost.

Lump-Sum Increases. A method of increasing the reinforcing properties of a merit increase system, the lump-sum increase offers the employee an opportunity to receive his salary increase in a nontraditional fashion. The standard practice requires that the merit increase be immediately incorporated into the employee's base salary, so that the full amount of the increase is received gradually over the course of the subsequent twelve months. Under the lump-sum method, this option is still available, but there are other options as well. The employee may choose to receive the entire annual increase all at once, in a lump sum. Or, he may choose to receive it in two semiannual or four quarterly installments. The company offering these options may or may not discount the lump-sum payments for the interest costs associated with the accelerated payment.

The lump-sum method increases the reinforcing properties of the merit increase system because it increases the visibility and impact of the raise in salary. A $2,400 salary increase carries greater impact if the employee receives the full amount in a single check as opposed to receiving an incremental $100 per semi-monthly pay check.

In addition to increasing the visibility of the merit increase, the lump-sum approach also increases quality of work life by enabling the employee to tailor his income stream to his own needs. A substantial amount of cash can be obtained to provide for the down payment on an automobile or to pay off a debt, for example.

A drawback of this approach is the additional administrative costs involved. The system must be administered, and the organization's payroll systems must accommodate the various options. A more serious concern is the nature of the merit system itself.

It is important that merit increases be truly performance-based, or the lump-sum approach will not achieve its aim for reinforcing performance. It will instead reinforce mediocrity, or whatever else drives merit increase practices.

The benefits and drawbacks of the various innovative approaches are summarized in Table 9-1.

Table 9-1
Advantages and Disadvantages
of Innovative Reward Systems

System	Advantages	Disadvantages
All-Salaried Work Force	Increases Trust Improves QWL	Higher compensation costs
Cafeteria Benefits	Meets individual needs	Administrative cost
Pay-for-Knowledge	Flexibility Personal growth	Training cost Higher salaries
Lump-Sum Increases	Meets individual needs Visibility of increase	Administrative cost

Gain Sharing

The most promising of the reward systems innovations, with such potential impact that it deserves more exhaustive treatment here, is gain sharing. Although technically not a recent innovation (its use dates back to the 1930s), this approach has enjoyed such rapid growth in popularity in recent years that it has the potential to dramatically alter the competitive landscape.

What is gain sharing? It is a group-based reward system that shares with employees, through a predetermined formula, the gains that accrue to the organization through improvements in productivity or some other performance variable. Employees, in other words, receive bonuses based on a predetermined measure (or measures) of organizational performance.

Profit-sharing, of course, has been a staple of reward systems for years and must technically be considered gain sharing. Profit-sharing traditionally, however, he served essentially as a retirement plan, with its payout deferred for many years. Because of this, and because of the fact that most employees have difficulty relating their efforts to the bottom line (especially in a large company), the traditional form of profit-sharing probably has little motivational value and is more properly viewed as a benefit (which of course has value in attracting and retaining people). This discussion of gain sharing will confine itself to those systems that provide a current cash payout (no less frequently than annually) and that are generally tied to a more micro measure than company profits.

Gain sharing plans come in a wide variety of shapes and forms. Some reward employees for physical productivity improvement, while others tie bonuses to broader measures of financial performance. Some plans reward only improvements in labor productivity, while others tie compensation to savings in materials usage and energy conservation. Some gain sharing programs share with employees the benefits of cost reduction, while others reward improvements in quality, service, or delivery performance. Some plans pay employees for improvements over historical performance levels, while others require that management-set targets be exceeded before bonuses are earned. There is no universally correct design for a gain sharing system; successful plans are tailored to the organization's needs and circumstances. The various design features are discussed in detail in Chapter 10.

What has led to the extraordinary growth of gain sharing in the 1980s? There are a number of reasons why companies are interested in this nontraditional form of reward:

Competitive Need. In this era of world competition, with Japanese wages half, and South Korea's 10%, of those paid in the United States, many companies have found themselves in an uncompetitive position. Under these circumstances, the organization can no longer live with the traditional practice of providing automatic annual increases in pay without regard to organizational performance. Gain sharing is a means of breaking this cycle; employees are able to earn higher pay through improved performance and the company can lessen its reliance on automatic increases. With compensation partly tied to organizational performance, the compensation system is more rational—pay increases as the company becomes more competitive.

Support for Employment Security. The rise of international competition and increased concerns about quality of work life have heightened the priority of employment security for both labor and management in recent years. As a result, many companies have sought means of ameliorating the pressures for work-force reductions during difficult times. Gain sharing serves this end, as it provides for a built-in financial adjustment during economic downturns. Compensation costs are partly variable with organizational performance, and these costs therefore decline along with the performance variable to which they are tied. With such an automatic adjustment, the need for layoffs is lessened or eliminated.

Trade-Off for Concessions. For the organization faced with the need to achieve immediate reductions in compensation costs, gain sharing can serve as a *quid pro quo* for wage concessions.

Motivational Benefits. Greater awareness on the part of management of the need for more effective productivity management practices has heightened concerns about inadequate reinforcement through the reward system. By sharing the gains realized by the organization through productivity improvement, increased motivation to improve performance is obtained and the reinforcing properties of the reward system are dramatically increased.

Support for Employee Involvement. The fact that the growth in the use of gain sharing has paralleled that of employee involvement is not purely coincidental. Since gain sharing is a group incentive, it reinforces teamwork and collaboration in problem-solving. In addition, by giving employees a "piece of the action," gain sharing provides a tangible reward for employee involvement in performance improvement and helps answer the inevitable question, "What's in it for me?" Gain sharing and employee involvement are highly compatible and mutually reinforcing systems.

Replacement for an Individual Incentive. Many companies are disillusioned and dissatisfied with existing piecework and other incentive systems that reward individual performance. The cost to maintain such systems is large, and they often lead to counterproductive behaviors in those organizations where interdependence among employees is high. In addition, they are characteristic of the rationalistic management style, which, as was suggested in Chapter 7, is giving way to a philosophy of management that places greater reliance on the human resource. Simply eliminating an individual incentive system is not practical, however, because of the inevitable resentment and alienation that will result on the part of those who have benefited from the system. Gain sharing offers a way out; employees have the opportunity to recover the lost incentive through improved organizational performance over time. To avoid immediate financial loss to employees, however, the company adopting this tactic generally must "buy out" the individual incentive (pay employees a lump sum equivalent to some period's worth of foregone incentive) or "red circle" the affected employees (adjust the base compensation to include their earned incentive) for a period of time.

Chapter 10 will discuss the design principles of gain sharing in greater detail.

REFERENCES

1. O'Dell, Carla, *Major Findings From People, Performance and Pay.* Houston, TX: American Productivity Center, 1986.
2. Gupta, Nina; Jenkins, G. Douglas, Jr.; and Curington, William P., "Paying for Knowledge: Myths and Realities," *National Productivity Review*, Spring, 1986.

10

Gain Sharing: Designed for Success

GAIN SHARING AS ORGANIZATIONAL CHANGE

As was suggested in the previous chapter, gain sharing represents an innovation that fits well with the concept of a productivity management process. Many of the same lessons learned about employee involvement apply to gain sharing as well. Like employee involvement, if gain sharing is implemented as another improvement technique, it will not likely succeed in the long run. Gain sharing represents a major change from traditional approaches to compensation and requires major shifts in organizational attitudes and supporting systems. And like employee involvement, the greatest value of gain sharing lies in its use in fostering organizational change. If gain sharing does not result in a more positive climate, improved labor-management relationships, a greater sense of shared objectives, and increased employee commitment, it is probably not worth the investment in time and resources that is required to design and support the system.

Gain sharing and employee involvement actually complement each other admirably. Employee involvement is the major mechanism for generating gains to share, and gain sharing is an excellent reinforcement for employee involvement. Many companies that have instituted gain sharing have done so, in fact, under the umbrella of employee involvement. Motorola, for example, views its gain sharing plans, which cover over 60,000 employees, not as separate systems but as integral elements of its participative management program.

Without employee involvement, gain sharing is unlikely to succeed. If a company installs a compensation system that rewards employees for improving organizational performance, but does not provide the means for employees to bring about those improvements, the predictable result is frustration and lack of employee support.

THE STANDARD GAIN SHARING PLANS

There is wide range of potential gain sharing designs, and it is vitally important that a gain sharing plan be custom-designed to fit the organization's needs and business situation. The various design components and options will be discussed later.

There exist several "standard" plans that are widely used in industry, and while their use without modification would be inconsistent with the just-expressed position that gain sharing plans should be tailored to the organization, it is nonetheless instructive to review these approaches.

The Scanlon© Plan

Developed in 1935, the Scanlon© Plan was the first of the gain sharing programs. Devised by Joseph Scanlon, a local union leader at the Empire Steel and Tinplate Company in Ohio, the plan was based on the premise that employees could make significant contributions to organizational performance if they were given the opportunity and support to do so. The initiative worked and was given credit for helping the company weather the depression. Word of the success spread, and the Scanlon© Plan was adopted by a number of companies in the 1930s and 1940s. The plan is still popular today and may be the most widely used of the standard plans.

An example of a typical Scanlon© formula is presented in Figure 10-1. Starting with net sales, an adjustment is made for inventory changes in or-

Net Sales	$ 1,500,000
Increase/(Decrease) in Inventory	300,000
Sales Value of Production	1,800,000
Allowed Payroll Costs (20%)	360,000
Actual Payroll Costs	312,000
Bonus Pool	48,000
Company Share (25%)	12,000
Employee Share	36,000
Deficit Reserve (25%)	9,000
Available for Immediate Distribution	27,000
Participating Payroll	$ 312,000
Bonus Percentage	8.7%

Figure 10-1. Example of Scanlon© formula.

der to arrive at sales value of production. This adjustment is made in order to relate labor costs to a relevant output, which in the case of a manufacturing facility, would be production rather than sales. A service organization, of course, would not have need for such an adjustment.

A percentage is then applied to sales value of production to yield "allowed payroll costs." This percentage (20% in the example presented) represents the actual average value of this variable during recent history (e.g., two years) and represents the baseline against which current performance will be gauged. If, as in the example, actual payroll costs are less than that "allowed," a gain has been achieved and a bonus pool results. The company share of this gain is deducted, leaving the employee share. A certain portion of the employee share is held back in a deficit reserve (this feature will be discussed shortly), and the remainder is available for immediate distribution to employees. This amount is divided into the participating payroll to obtain the bonus percentage. In the example presented, each employee would receive a bonus check equal to 8.7% of his wages or salary for the period in question.

The Scanlon© Plan clearly rewards employees for improvements in labor productivity, as measured by the ratio of payroll to value of production. A careful analysis of the example presented in Figure 10-1 reveals another interesting feature of this approach: It also rewards employees for financial gains resulting from increases in selling prices. While it certainly could be argued that employees have no control over the market conditions that determine selling prices, many companies embrace this type of formula because it ties compensation to broader determinants of organizational performance than labor productivity alone and helps promote the philosophy that "We're all in this together."

The Rucker© Plan

Developed in the late 1930s by the economist Alan Rucker, the Rucker© Plan is similar to the Scanlon© Plan in many respects: it is based on a people-oriented management philosophy and it relates payroll costs to financial measures of output. An example of the Rucker© formula is presented in Figure 10-2.

The primary distinction between the Scanlon© and Rucker© plans is that the latter relates payroll costs to "value added" rather than to sales value of production. Value added is calculated by subtracting from sales the total of all outside purchases of materials and services. A manufacturing organization purchases raw materials, supplies, energy, and contract services. These inputs enter the productive process, where labor and capital are applied to produce an output that has greater value to the consumer than the total

Sales Value of Production	$ 1,800,000
Outside Purchases:	
Raw Materials	800,000
Energy	200,000
Contract Services	100,000
	1,100,000
Value Added	700,000
Rucker Standard (48%)	336,000
Actual Payroll Costs	312,000
Bonus Pool	24,000
Deficit Reserve (25%)	6,000
Available for Immediate Distribution	18,000
Participating Payroll	$ 312,000
Bonus Percentage	5.8%

Figure 10-2. Example of Rucker$^{©}$ formula.

value of the purchased materials and services that were utilized as inputs. The organization has therefore "added value" to these purchases.

In the Rucker$^{©}$ Plan, actual payroll costs are compared to the historical relationship of payroll to value added (the Rucker$^{©}$ standard) for the organization. Any improvement over this historical relationship represents a gain and results in an employee bonus. Unlike the Scanlon$^{©}$ Plan, there is no company share deducted from the bonus pool. If the gain is the result of an increase in value added, the company share is actually the inverse of the Rucker$^{©}$ standard. If, on the other hand, the gain results from lower labor costs on unchanged value added, the employee share is effectively 100%.

The major attraction of the Rucker$^{©}$ Plan is that it rewards employees for savings in material and energy usage as well as for labor productivity improvements. Reductions in scrap losses and energy usage reduce the amount of purchases of these items and thus increase value added. This in turn increases the amount of allowed labor costs and, assuming that actual payroll costs remain unchanged, generates a gain. Rucker$^{©}$ thus has appeal to manufacturing organizations that seek to improve materials and energy usage through employee efforts.

IMPROSHARE$^{©}$

The third of the standard gain sharing plans is very different from the Scanlon$^{©}$ and Rucker$^{©}$ plans and appeals to managers who are not comfortable with the financial-based measures characteristic of these plans. They prefer to tie compensation to physical measures of productivity and thus eliminate the impact of changes in selling prices and developments that are beyond the control of employees.

Developed by an industrial engineer named Mitchell Fein, IMPROSHARE© (an acronym for Improved Productivity through Sharing) rewards employees for improvements in physical productivity. An example of an IMPROSHARE© calculation is presented in Figure 10-3.

	Production In Units	Direct Labor Hours/Unit
Product A	400	5
Product B	600	3
Product C	300	2

Formula Calculation:

A: 400 units × 5 hours × 2.15	4,300
B: 600 units × 3 hours × 2.15	3,870
C: 300 units × 2 hours × 2.15	1,290
TOTAL IMPROSHARE ≤ HOURS	9,460
Actual Hours	8,790
Gain	670
Company Share (50%)	335
Employee Share	335
Actual Hours	8,790
Bonus Percentage	3.8%

Figure 10-3. Example of IMPROSHARE© calculation.

The IMPROSHARE© calculation is based on direct labor standards. During the measurement period, units of production for each product are multiplied by the appropriate direct labor standard, stated in hours. In the example shown, products A, B, and C have direct labor standards of five, three, and two hours per unit, respectively. These are not necessarily engineered or target standards, but may be based on historical performance.

The results of these calculations are then multiplied by a fixed factor (2.15 in the example) in order to provide a baseline that encompasses all of the organization's labor, indirect as well as direct. In other words, total labor hours expended in our hypothetical organization have historically been 2.15 times the direct labor hours utilized.

The aggregate value of these calculations (IMPROSHARE© hours) is compared to actual hours worked during the period, and, to the extent that actual hours are less, a gain has been realized. Half of the gain accrues to the benefit of the company, and the other half is divided into total hours worked to derive a bonus percentage. Bonuses are typically calculated and paid weekly.

Significantly, dollars never enter the picture in calculating the gain. Bonuses are thus based on physical productivity and are not directly affected by marketplace and other external developments.

IMPROSHARE© also varies from the Scanlon© and Rucker© plans in that bonus payments are capped—the payout cannot exceed 30% of wages. This feature adds additional complications, requiring the use of a deferral account and buy-back provisions. The mechanics of these features will be discussed in the sections ahead.

IMPROSHARE© also differs from the other standard plans in additional ways. It does not utilize a deficit reserve, for example, and it does not prescribe a specific approach to employee involvement.

DESIGNING A GAIN SHARING SYSTEM

A review of the standard gain sharing plans illustrates an important point: There is a wide range of possible designs for a gain sharing system. The potential options are actually far greater than those represented by the standard plans, as we will see shortly. Some of the most successful gain sharing plans, in fact, are quite unique and do not resemble any of the standard plans.

Installing, without modification, a standard or predetermined gain sharing system may be a mistake. Every organization is different, and the management philosophy and objectives for gain sharing will vary from one organization to another. A particular design that is highly successful in one company may well fail in other companies.

Organizations are well-advised, therefore, to design their gain sharing system from scratch. Management must ensure that the plan is consistent with management philosophy, strategic objectives, organizational culture, technical processes, management style, employee capabilities, accounting systems, communications practices, marketplace conditions, and existing improvement opportunities. There is nothing sacred about the standard gain sharing plans, and an attempt to force-fit a predesigned system will be no more likely to succeed than an inappropriate employee involvement or productivity improvement technique.

Since gain sharing plans should be custom-designed, it is important to be aware of the various design components and the options available for each. The rest of this chapter will focus on these design components.

CRITICALITY OF DESIGN

A poorly designed gain sharing system will, at best, limit the potential benefits and may well result in an outright failure. There exists an enormous range of successful gain sharing system designs, and it is foolhardy to risk failure by attempting to force-fit one of the small number of standard plans into the organization. As was suggested earlier, some of the most successful gain sharing programs look nothing like any of the standard plans, but are unique in their design.

All gain sharing systems do have certain components around which design decisions must be made. These components are:

1. Designation of the *group* to which the gain sharing plan applies.
2. A *formula* through which group performance is measured.
3. A *baseline* against which improvements in the formula are compared.
4. A *share* arrangement, or basis for dividing the gains between the organization and its employees.
5. A *payout frequency*.
6. A *payout distribution* method to allocate the employee share of gains to individual employees.

There are three additional design features which are not technically necessary in order to have a gain sharing system, but are highly desirable:

1. A *deficit reserve* to reduce to some degree the risk to the organization.
2. *Employee involvement* structures to provide the means for employees to bring about gains.
3. A means of adjusting the formula or baseline for major *capital investments* made by the organization.

Finally, an optional feature that is sometimes found in gain sharing systems is a *cap* on the payout. Where this feature exists, there is normally an associated *buy-back* provision.

It must be stressed again that there is no "right" way to address the various design components. There are certain principles that must be considered in building the design, but these are largely based on common sense and can be observed with a wide range of options.

The design criteria and common options for each of the components are discussed in detail below.

The Group

The first question that must be asked when designing a group incentive is, "Who is the group?" None of the other design components can be rationally addressed until this question is answered.

There are three basic options for defining the boundaries of participation in the gain sharing plan:

Non-Exempt Only. Some organizations choose to limit participation in the gain sharing plan to the nonexempt or hourly work force. The rationale for this decision is that the exempt or salaried work force benefits from other motivational systems (merit increases, promotional opportunities), while the hourly work force receives only general or contractual increases and thus lacks motivation to improve performance.

A complicating factor here is the first-line supervisor. The supervisor certainly affects the performance of the hourly worker and often relates more closely to the hourly work force than to management. For these reasons, some organizations choose to include first-line supervisors in the plan.

The major negative to this approach is that it may aggravate an existing problem: the disaffection that often results from the artificial line of demarcation between the salaried and hourly work force. As was suggested earlier, many organizations have concluded that this distinction between groups of employees is a major cause of low commitment and poor performance. Another potential problem is pay compression; if the program is successful, compensation for hourly workers may increase faster than for salaried employees.

Site Inclusive. The most common decision on the "group" issue is to include everyone (except, possibly, those covered by an executive performance bonus) at a given site or organizational unit, such as a plant or a division.

The obvious appeal of this approach is that it encourages a sense of common objectives among all employees and fosters teamwork and collaboration in improving performance.

Multiple Product/Process Approach. While management may wish to include all employees in a given organizational unit's gain sharing plan, the sheer size or diversity of the unit may argue against a single plan or formula. A division of an aerospace firm, for example, was legitimately concerned about the effectiveness of any plan that would cover its 25,000 people under a single formula.

A solution is to develop multiple plans, each with its own formula, designed to reward the type of performance that is relevant to various natural work groups within the organization. These subgroups may be defined around products (or services), processes, or functions. An example of this approach is found in Nucor Corporation, a steel producer, which has separate plans for production employees, department heads, indirect employees, and senior officers. Another company's steel mill has separate plans for its two major production processes, which operate largely independently of one another.

Several variations on this option are available. Each defined subgroup may have its own bonus pool, tied to improvements in its specific formula. Or there may be a single bonus pool, shared by all, which represents the aggregate improvement in all of the formulas. Variations incorporating both of these options are even found; the bonus of each subgroup, for example, may be dependent in part on that group's performance and in part on the aggregate performance of all groups.

A drawback of this approach is the potential for destructive competition between groups that are interdependent to some degree. Dissatisfaction may also arise if one group considerably outperforms another.

The Formula

Once the organizational boundaries of the plan have been defined, it is natural to consider the formula. If gains are to be shared with the work force, a formula to measure those gains is an obvious necessity. While the variety of formulas found in gain sharing plans is enormous, we can broadly categorize them into three types: physical formulas, financial formulas, and families of measures.

Physical Formulas. Corresponding most closely to the traditional definition of productivity, physical formulas typically reward employees for improving the relationship between physical units of output and physical units of input. Nucor, for example, ties the bonus for its melt shop employees to tons produced per operating hour.

Physically-based plans in complex, multiproduct organizations are often based on standards so that the effects of changes in product mix are automatically factored into the calculation. The use of standards (or some other weighting mechanism) ensures that a shift in mix toward a low labor-content product will not result in an artificial improvement in a units-per-man-hour formula.

The best known of the physically-based gain sharing plans is IMPROSHARE©, described earlier. The IMPROSHARE© formula basically compares actual hours worked with standard direct labor hours earned, adjusted upward by the indirect/direct labor ratio. Half of the resulting percentage improvement translates into an employee bonus of like percentage of wages. A prerequisite for the use of IMPROSHARE© is, of course, a process that is susceptible to the use of production standards.

Companies that choose a physical formula as the basis for calculating gains generally do so because they wish to reward employees for those aspects of organizational performance that they most directly control. Physical formulas are not affected (except indirectly) by changes in selling prices or market conditions. The use of a physical formula does raise the possibility, however, of paying bonuses during periods of declining or nonexistent profitability, as weak markets could occur in the face of rising productivity.

Financial Formulas. Those organizations that wish to tie employee bonuses more closely to overall organizational performance generally choose a financial formula as the basis for their gain sharing plan. While a financial formula may still relate output to input, it does so in dollar terms, as in the ratio of sales to payroll.

The use of a dollar-denominated formula is not simply a minor variation of the physical formula approach; it has major implications for the nature and scope of the gain sharing plan. With a financial formula, employees' bonuses are affected by pricing decisions and marketplace conditions, as well as by physical productivity performance. For this reason, companies that select financial formulas are usually motivated by a desire to create more of a "common fate" orientation within the organization. We're all in this together, they reason, and employees' compensation should therefore rise and fall in concert with overall organizational performance.

The oldest of the traditional gain sharing plans—the Scanlon© Plan and the Rucker© Plan—are based on financial formulas. As indicated earlier, the Scanlon© Plan rewards employees for improving the ratio of payroll costs to sales value of production (sales adjusted for inventory changes) while the Rucker© Plan bases bonuses on improvements in the ratio of payroll costs to value added (sales value of production less purchased materials and services). As with any financially based system, bonuses generated under these plans are affected by changes in selling price and product mix as well as by productivity improvements.

The Rucker© formula, as was noted previously, inserts an additional element into the gain sharing picture. Since gains are measured in terms of improvements in value added, bonuses may be earned by employees through reductions in materials and energy usage as well as through improvements in labor productivity.

In view of the potential impact that employees have on a variety of costs in most organizations, multicost gain sharing plans are rapidly gaining favor. The focus of this approach is on improving the ratio of sales to a defined list of cost items, up to the total operating costs of the unit. Eggers Industries, a wood products manufacturer, includes thirty cost elements, which aggregated to 81.5 percent of sales at the time of program start-up, in its gain sharing plan. After four years of operation, Eggers' program has reduced those thirty cost elements to 69.6 percent of sales.[1]

The ultimate financial plan, of course, is profit sharing. While profit sharing has been in existence for over a century, its use has traditionally been associated with a retirement plan and has been based on the bottom line of a large organization. While potentially valuable as a benefit, the traditional approach to profit sharing is of questionable value as a motivator and a compensation tool. After all, how motivated can an employee be to improve his performance if he is one of 40,000 and the payoff is 30 years hence? In recent years, however, there has been a noteworthy trend toward the use of profit sharing in smaller organizational units and with a current (i.e., annual) payout.

Family of Measures. An increasingly popular approach does not fit either of the preceding categories, which are characterized by a single measure of performance.

The family-of-measures approach is employed by those organizations that wish to focus employees' efforts on multiple performance variables with separate indicators for each variable. Several different indicators are selected, and improvements in those indicators are aggregated to create a bonus pool.

An excellent example of this approach is provided by a Mobay Corporation plant in New Martinsville, West Virginia. In one recent year, the plant's gain sharing program was structured around ten indicators:

1. Tools and Safety Supply Usage: actual usage versus prior three-year average.
2. Salvage and Reuse Savings: 50 percent of price of items salvaged.
3. Energy Usage: volume-adjusted usage compared to prior year, less savings attributable to capital investment.
4. General Supply Usage: actual usage versus prior year.
5. Ferrous Oxide Reactor Time Saving: gross reactor cycle time less base period gross reactor maintenance cycle time.
6. Ferrous Oxide Milling Recycle Reduction: current year performance compared with goal.
7. Wastewater Treatment Cost Savings: volume-adjusted performance versus prior year.
8. ECD Solid Waste: volume-adjusted performance versus prior year.
9. Polyurethane Area I Error Reduction: $10,000 savings per incident less than base of 24 incidents per year.
10. Polyurethane Area II Composite Performance: savings associated with composite measure, including quality, housekeeping, and organic losses.

The aggregate savings realized from these ten indicators creates the bonus pool, which is shared 50/50 with employees.

The Mobay indicators are reviewed annually each year by a joint labor-management committee, which determines what indicators will form the basis for the new year's gain sharing program. Some indicators may be retained from year-to-year, while others will be discarded and replaced by new indicators.

Perhaps the most comprehensive use of the family-of-measures approach is found at Motorola, whose gain sharing plans cover 60,000 employees on a unit-by-unit basis. In its manufacturing units, Motorola's plan is based on several performance indicators, including production cost, quality, delivery, inventory levels, and safety. Other sets of measures apply to administrative and technical groups.

The Baseline

Once the group has been delineated and a formula has been developed to measure improvement, the obvious question is, "Improvement over what?" What is the baseline against which improvement will be measured and rewarded?

There are two basic options with many variations. The first option—a historical baseline—is employed by most of the standard plans, such as Scanlon©, Rucker©, and IMPROSHARE©, and will probably be found in the majority of existing gain sharing programs. The average level of the chosen formula for a past period, typically ranging from six months to three years, is established as the baseline. Any improvement over this average past performance results in a gain to be shared.

In some cases, however, a historical baseline simply is not appropriate. Perhaps the historical data have been rendered invalid by recent changes in technology, processes, or products. Or perhaps the company is presently uncompetitive and must achieve higher productivity levels before it can consider raising compensation.

Where history does not provide an acceptable baseline, the remaining option is a targeted baseline. The baselines for Motorola's multiple measures, for example, are goals that are set annually; shared gains occur only when the goals are exceeded.

A related design issue is the frequency of changes in the baseline. At one extreme is the fixed baseline found in the Scanlon© and Rucker© plans. The baseline remains frozen (at least until compelling reasons dictate that it change), and employee bonuses therefore reflect improvements made in previous years as well as in the current period. At the other extreme is the baseline that ratchets up each year to the level of performance achieved in the previous year. With this option, employees must generate ever increasing improvements in order to continue earning bonuses over time.

There are many variations between the two extremes. A common approach, for example, is the rolling baseline: Each year the baseline is recalculated as the average performance for some specified number of previous periods. Employees do not immediately lose the benefits of the prior year's performance, but the gains do roll out of the system over time. As an example, the gain sharing plan in use at some plants of the 3M Company utilize a rolling twelve-quarter baseline.

The Share

With the group identified and a formula and baseline defined, the organization can now calculate the performance gains that are to be shared with the workforce. But what is to be the sharing arrangement? Should it be 50/50 or some other ratio?

Equity is the ultimate criterion for the sharing decision. Given the nature of the business, the nature of the formula, and other considerations, what constitutes a fair and equitable distribution of the gains? Several considerations affect this decision.

If the formula measures only labor productivity, a higher percentage of the gain is typically paid to employees than in a plan based on a multicost formula. The Scanlon© Plan, as one example of a labor-only formula, traditionally returns 75 percent of the gain to employees. It is not unheard of, in fact, for labor-only plans to award employees 100 percent of the calculated savings. On the other hand, as more cost items are included in the formula, the plan begins to approach a profit-sharing arrangement, and the employee share generally declines. Typical multicost plans (short of true profit sharing) pay out 25 to 50 percent to employees. An example is the multicost plan of American Valve and Hydrant Manufacturing Company in Beaumont, Texas, which pays employees 32 percent of the achieved gains.

Another consideration is the capital intensity of the business. Given their substantial capital investment needs, capital-intensive businesses tend to retain a larger proportion of gains than those companies with lesser investment requirements.

The frequency with which the baseline is changed also affects the sharing decision. If the baseline changes as performance improves, requiring ever-greater productivity improvement in order to generate a bonus, employees should receive a greater share of the gains than might be appropriate under a fixed-baseline approach, where employees continue to earn bonuses based on previous years' gains.

The motivational impact must also be considered. One could expect that employees would be more inclined to support a plan that awarded them 75 percent of the gain rather than 25 percent.

An unusual approach to the share component is taken by St. Luke's Hospital in Kansas City; the employees' share of gains (calculated department by department) ranges from zero to two thirds, depending on the degree to which the hospital exceeds its overall financial goals.

Payout Frequency

The frequency of bonus payment for most gain sharing plans is based on natural calendar divisions: weekly, monthly, quarterly, or annually.

The first criterion for the frequency decision is data availability; an organization obviously cannot pay gains on a weekly basis if the data to calculate the gain are only available monthly.

Beyond the data availability consideration, there are trade-offs to consider. More frequent payouts, for example, provide more frequent reinforcement for performance improvement but raise administrative costs and lower the magnitude of each bonus check. The benefits of frequent reinforcement may be largely lost if payouts are small. On the other hand, large bonuses received at very long intervals may have less than optimum impact on employee behavior as well.

A final consideration is the inherent variability of the formula itself. If the selected formula has a history of substantial variation from period to period, payouts should be at intervals long enough to smooth out the ups and downs and therefore ensure some stability in bonuses.

Payout Distribution

How is the employees' share of any gains to be divided among the employees? There are three basic options for this feature:

☐ Percent of income—the bonus pool is translated into a percentage of salaries and wages, with each employee receiving an equal percentage of his compensation.
☐ Equal shares—every employee receives the same absolute dollar amount.
☐ Hours worked—the bonus is stated in terms of dollars per hour worked and applied to individual employees accordingly.

The percent-of-income approach is utilized in about 50 percent of the plans, with other approaches used in the remainder. As with other design features, variations on the basic options exist. Eggers Industries, for example, uses the percent-of-income approach with the added proviso that no salaried employee may receive a larger payment than the highest bonus earned by an hourly employee.

Deficit Reserve

While not an absolute requirement for a gain sharing plan, a deficit reserve reduces the company's risk somewhat and is a widely utilized feature.

A portion of the employees' share of each payout (typically ranging from 15 to 30 percent) is withheld from employees and placed in a reserve account. For any period in which organizational performance is negative relative to the baseline, the company reclaims from that reserve an amount equal to employees' share of that deficit. The reserve account can carry a negative, as well as a positive, balance. At the end of each year, the reserve is zeroed out by the company: If the account is positive, the balance is paid out to employees; if the balance is negative, it is absorbed by the company.

The use of a deficit reserve protects the company against an undesirable situation: paying higher compensation during a year in which overall performance, as measured by the gain sharing formula, is negative. This scenario could easily occur without a reserve, for even in the worst of years, there may well be some months in which a bonus is earned. Since base compensation is still paid in the negative months, the net effect would be higher compensation costs for the year.

Those organizations that choose not to have a deficit reserve can reduce this risk by basing the bonus on the average of several periods' performance. IMPROSHARE©'s weekly payout, for example, is typically based on the average calculated gain for several past weeks. Periods of negative performance may then offset gains earned in other periods.

Employee Involvement

Like the deficit reserve, supporting employee involvement structures are not a mandatory feature of a gain sharing plan. However, the likelihood of success without them is slim.

In traditionally managed (i.e., autocratic) organizations, the opportunity for employees to contribute to performance improvement is limited. Supervisors are not highly receptive to employee ideas, and effective structures to stimulate, gather, and evaluate ideas for improvement are absent. When a gain sharing program is implemented in this type of environment, employee frustration is the predictable result. Employees are provided the opportunity to earn more money through productivity improvement, but their efforts to bring about improvement are not supported.

Employee involvement mechanisms, therefore, represent a vital supporting feature for any gain sharing plan. As was discussed in Chapter 8, there are a multitude of employee involvement structures, ranging from suggestion systems to self-managed work teams, and the organization should carefully select those techniques that will be most effective in its environment.

An involvement structure that is often associated with gain sharing programs, particularly the Scanlon© Plan, is a committee-based suggestion system (Figure 10-4).

This approach utilizes departmental suggestion committees to evaluate employee ideas for improvement. These committees typically include the department supervisor and two or three nonmanagement employees who are elected by their peers.

These committees have the authority to implement, at their own discretion, those ideas that meet two criteria: 1) They do not affect other areas of the organization, and 2) Their cost of implementation is less than some maximum amount, generally $200 to $500.

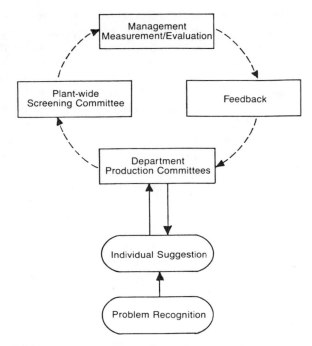

Figure 10-4. Committee-based suggestion systems.

Those ideas that cannot be approved by the departmental committees are referred to the steering committee that manages the gain sharing plan. This committee usually consists of both managers and nonmanagement representatives. The steering committee evaluates the idea in question, reaches a decision, and provides feedback to the referring departmental committee, which in turn provides feedback to the individual suggestor.

Whatever involvement structures are utilized, time and resources spent to provide these mechanisms will earn an excellent return in the form of a more viable gain sharing system.

Capital Investment

At some point during management's exploration of gain sharing, a question involving the treatment of gains resulting from capital investment invariably arises. Since capital investment decisions are usually initiated by management and require the commitment of invested or borrowed capital, management is naturally reluctant to share these gains with the workforce. Furthermore, the sharing of these gains might invalidate the financial justification upon which the investment decision was made.

On the other hand, employees do influence the effectiveness of any capital investment implementation. They can enthusiastically cooperate in bringing the new equipment up to its peak performance capability, or they can resist the implementation effort and thus delay or reduce the realization of the intended gains. Furthermore, those organizations seeking to foster greater teamwork or to create a "common fate" orientation have difficulty justifying the total exclusion of employees from the enjoyment of the gains associated with capital investments.

In practice, most organizations do adjust the plan for capital investments above a given investment threshold, such as $50,000; it is not desirable to continually adjust the plan for the frequent, minor investments that most organizations make. In terms of the nature of the adjustment, there are several options:

Full Adjustment with Lag. Under this approach, the baseline is adjusted for the full impact of the capital investment. Employees remain whole, as they continue to receive bonuses based on the gains that had been realized prior to the capital investment. This option is graphically depicted in Figure 10-5.

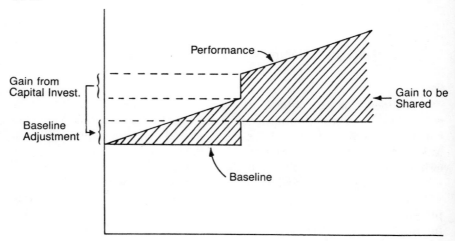

Figure 10-5. Full capital investment adjustment.

Organizations selecting this option generally delay the adjustment for six months to one year following the installation of the new equipment. This lag provides a motivational benefit, as employee support for timely installation will result in the related gains entering the bonus pool until the baseline adjustment is made. The lag in adjusting the baseline also enables management to obtain a reasonably accurate fix on the true gains, as opposed to the theoretical gains found in the financial justification for the investment.

Partial Adjustment. In the interests of promoting the common-fate philosophy, many organizations choose to make only a partial adjustment for the gains associated with capital investments. IMPROSHARE©, for example, was designed to provide for an 80 percent adjustment to the baseline for capital-related gains. This approach, which may be applied with or without a lag, ensures that employees at least partially benefit from management's capital investment decisions.

Payback, Then Share. Yet another option is to fully adjust the baseline immediately (or as soon as the new equipment is producing up to its capability), and then return to the unadjusted level when the gains have payed back the amount of the investment. Management thus ensures that the capital investment is recovered prior to employees sharing the gains.

No Adjustment. The major drawback of all of the preceding options is complexity. Management must explain and justify to employees the adjustments made or risk losing their commitment to the plan. This can indeed be a formidable task, given the arcane financial analyses typically employed to evaluate capital investment proposals. In addition, any calculation of gains is likely to be based on estimates, and to the degree that actual results vary from those estimated, the baseline adjustment will be inaccurate and possibly unfair to employees.

In view of the difficulties associated with capital investment adjustments, some organizations choose to make no adjustments, but to deal with the problem in other ways. The use of a rolling baseline, for example, will ensure that capital-related gains are shared with employees for only a limited period of time. The use of caps and buy-backs (see following section) also ameliorate the problem, as major capital investments (or the cumulative effect of several investments) produce gains that exceed the cap and trigger an automatic adjustment to the baseline, with compensation to employees.

Finally, there are examples of gain sharing programs where management simply decides to share the gains from capital investments because of a desire to promote the common fate philosophy.

Caps and Buy-Backs

A feature found in IMPROSHARE© and some custom-designed gain sharing plans is a capped payout. A cap limits the amount of bonus that can be paid to employees to a certain percentage of income (30 percent in the case of IMPROSHARE©).

A cap does not mean that employees cannot benefit from further gains beyond the limit imposed; if that were so, the gain sharing program would lose its motivational advantages once the cap had been reached. Rather, the employee share of gains in excess of the cap are credited to a deferral account and are paid out subsequently during periods when bonuses fall below the prescribed maximum. Rather than being lost, the excess gains are simply deferred to a later time.

When organizational performance consistently exceeds the cap, however, the company has a problem. The balance in the deferral account grows and grows, as there are few below-cap periods that will allow the deferred amounts to be paid out. The solution to this problem is the buy-back, a feature that normally accompanies a capped bonus.

The buy-back works like this: the baseline is adjusted so that the bonus earned at the current performance level falls somewhat below the cap. Since this action reduces bonuses, employees are compensated by the payment of a lump sum equal to one year's worth (or some other period) of the foregone bonus. Since payouts no longer exceed the cap, the balance in the deferral account will now begin to be paid out.

An interesting variation of the cap and buy-back approach, which eliminated the need for a deferral account, was employed by a Florida Steel plant in Jackson, Tennessee. A 25 percent cap was established, but gains above the cap are paid out in a normal fashion. When gains exceed the cap for six consecutive months, a buy-back is triggered. This approach would more properly be called a "trigger point," rather than a cap, since it does not place a ceiling upon employee bonuses.

The cap and buy-back features serve at least one important purpose: they ensure that the gain sharing bonus will not account for an inordinately large proportion of employees' compensation. There are examples of gain sharing plans that have paid bonuses in excess of 100 percent of wages, and, while this level of payout certainly is indicative of a successful program, it does pose a potential problem. If the bonus comes to account for a large proportion of compensation on an on-going basis, employees may suffer during economic downturns or other extended periods of low organizational performance. Employees receiving regular and recurring bonuses of high magnitude may adjust their standard of living and financial commitments accordingly, with disastrous consequences when these bonuses decline significantly or disappear altogether.

DESIGNED FOR SUCCESS

The foregoing review of gain sharing design components is far from exhaustive; the options presented here are those most commonly seen or were based on broad categorizations. The manner in which the standard gain sharing plans—Scanlon©, Rucker©, and IMPROSHARE©—address the major design components is summarized in Tables 10-1 through 10-3.

Most plans, if they are carefully designed to fit the organizational style and circumstances, vary from the "standard" approaches in significant ways. The lesson is that there is no "best" approach to any of the compo-

Table 10-1
The Scanlon© Plan
Design Features

Group	All Employees
Formula	Traditionally, the ratio of sales value of production to payroll costs
Baseline	Historical
Share	Traditionally, 75 percent to employees
Payout Frequency	Monthly
Employee Split	Percent of income
Deficit Reserve	Yes
Employee Involvement	Departmental suggestion committees

Table 10-2
The Rucker© Plan
Design Features

Group	All employees
Formula	Ratio of value added to payroll costs
Baseline	Historical
Share	Variable, depending on historical value added/payroll ratio and on type of gain
Payout Frequency	Monthly
Employee Split	Percent of income
Deficit Reserve	Yes
Employee Involvement	Plant-wide suggestion committee

Table 10-3
IMPROSHARE©
Design Features

Group	All employees or hourly employees only
Formula	Comparison of actual hours worked with standard direct labor hours earned, adjusted for indirect labor
Baseline	Historical or target
Share	50/50
Payout Frequency	Weekly
Employee Split	Percent of income
Deficit Reserve	No
Employee Involvement	Not specified

nents of a gain sharing system. A plan that is tailored to the needs and requirements of the organization offers the greatest opportunity for long-term success.

Given the importance of a custom-designed gain sharing plan, the design process itself becomes a key issue. Many organizations constitute a multifunctional and multilevel design team that is charged with considering the various design components and developing a detailed gain sharing plan design. Typically, the design team is first provided basic education about gain sharing and its components, and then meets periodically to construct the system. Top management may well provide guidance on the major decisions (such as the definition of the group) and reviews the design team's deliberations as they proceed. This ensures that the plan that ultimately emerges from this effort will not contain any surprises or features that are unacceptable to senior management.

The more progressive organizations will include hourly employees or union leaders on the design team. The obvious benefit of doing so is a greater level of employee and union commitment to the resulting plan. If lower-level employees are not involved in the design process, management will likely have to "sell" the plan to the workforce. Some companies even employ rather elaborate structures, such as departmental subcommittees to which the individual design team members report following each meeting.

Organizations should be willing to allow six months to one year for the design of a tailored gain sharing plan. This will ensure that the design components are thoroughly and systematically evaluated and that the system will "fit." The time invested in a careful and deliberate design process will pay off later in the avoidance of "fixing up" time and in the effectiveness of the system.

REFERENCE

1. *Labor-Management Cooperation Brief: Employee Involvement and Gain Sharing Produce Dramatic Results at Eggers Industries.* Washington, DC: U.S. Department of Labor, Bureau of Labor-Management Relations and Cooperative Programs, 1985.

11

The Quality Connection

WHAT IS THE CONNECTION?

How does quality relate to productivity? Do these performance variables reinforce each other or are they mutually exclusive? Must improved quality come at the expense of productivity?

Management traditionally has viewed quality and productivity essentially as trade-offs. To achieve significant improvements in one, some degradation in the other must be accepted. Quality could only be improved at the expense of productivity, and vice versa. Yet many firms today operate under the philosophy that improved quality *results* in improved productivity. How are these incompatible viewpoints reconciled?

The problem lies, in part, in definition. If quality is viewed in an absolute sense—improved quality equating with absolute goodness or ever-tighter specifications—it may indeed be difficult to envision how improved quality can be achieved without added cost. If, on the other hand, quality is viewed as *conformance to specifications* (a definition widely accepted by business today), the relationship to productivity becomes more apparent. If the product or service is produced with defects (fails to meet specifications), then it must be reworked, reprocessed, or reproduced. The result is more resources—people, materials, equipment—required to produce a given amount of product or services meeting specifications.

This leads us to the concept of *process* quality, which has a clear and direct correlation with productivity. While an organization's finished products (or delivered services) may ultimately conform to specifications, the quality of the process that produced those products or services can vary widely and will have a major bearing on the productivity of that organization. If substantial amounts of product must be reworked or reprocessed, if raw materials are defective, if waste and materials losses are excessive, if scrap losses are high, the organization can hardly lay claim to high levels of either quality or productivity.

Poor quality performance increases the inputs required to produce a given amount of good output. Rework certainly increases the amount of labor required and probably increases capital, materials, and energy inputs as well. Waste and scrap losses clearly increase the materials required for a given level of production, not to mention the labor, equipment and space required to handle these losses. And poor quality performance increases the need for inspection and controls, which of course require added resources.

With poor quality, a substantial amount of an organization's resources must be devoted to correcting defects and handling waste rather than producing products or services. As quality improves then, the resources required to produce a given amount of output decline, and that translates to improved productivity.

The experience of the Computer Systems Division of Hewlett-Packard is instructive.[1] Printed circuit boards were assembled by the division utilizing a kit of parts gathered together in the stores area. The parts were inserted, loaded, and soldered onto boards in lot sizes of 20 to 200.

On average, 98 percent of the kit parts were available in the stores area when work orders were issued. However, the kits required a large number of parts (as many as 100), and, in spite of the 98 percent in-stock level, a large percentage of the kits pulled had one or more parts on back order. These kits were nonetheless delivered to production, and the assembly process proceeded as far as possible. The partially completed boards were then pulled off the production line and stored on shelves until the missing parts arrived. This practice was followed because of management's belief that it was important to keep people working even if the entire assembly process could not be completed.

The assembly manager undertook an experiment in 1983; he ordered that kits not be delivered to the assembly area until all parts were available. The experiment proceeded despite concerns by others that production schedules could not be met without building the partial assemblies as far as possible.

Indeed, the output of the assembly area initially slowed as partial assemblies were completed and little new material entered the department. Employees' idle time increased dramatically.

By the fourth week of the experiment, however, the pipeline had filled with complete kits and production had returned to the levels maintained prior to the change. But workers still were often idle, as the amount of labor required to assemble a board had been essentially reduced by half. It was apparent that almost half of the time of the assembly department staff had been devoted to handling unfinished work-in-process, moving material onto and off of the storage shelves, and expediting the partially completed assemblies. Under the new practice, none of these tasks were necessary.

In addition, the new procedure exposed a high variability in the workload, a condition that had been hidden by the substantial volume of unfinished work-in-process. At certain times there was very little work in the assembly area, while at other times there was far more work than could be handled. The solution to this problem was to reduce lot sizes, so that smaller lots were delivered to the assembly at frequent intervals.

The effect of the smaller lot sizes, together with the elimination of handling partially-finished assemblies, resulted in an almost two-thirds reduction in manufacturing cycle time.

Through the simple step of improving the quality of incoming parts kits, significant improvements in productivity were realized. The productivity-quality connection had been demonstrated.

MANAGING QUALITY

The principles of quality management have undergone a considerable transformation in recent years. Thanks largely to the efforts of such individuals as W. Edwards Deming, J. M. Juran, and Philip Crosby, American managers are beginning to view the quality issue in a different light. The contemporary thinking on quality includes the following assumptions:

Poor Quality Costs More than Is Traditionally Recognized. The *cost of quality*, a concept popularized by Philip Crosby[2], is routinely calculated by a number of firms and is often found to exceed 20 percent of revenues. Crosby defines cost of quality as *the expense of nonconformance to specifications* and divides these costs into three categories:

1. Prevention costs are those costs incurred to prevent defects and errors in the development and production of a product or service. These costs include such items as design reviews, drawing checking, supplier evaluations, specification reviews, quality audits, and preventive maintenance.
2. Appraisal costs are incurred in inspecting or evaluating products or services to determine whether they conform to requirements. Examples of appraisal costs include receiving inspection, supplier surveillance, product acceptance, final inspection, and status reporting.
3. Failure costs are those costs that result from the nonconformance of a product or service to requirements. These costs include engineering change orders, product redesign, purchasing change orders, scrap, rework, warranty charges, and product liability costs.

Quality Cannot Be Inspected into the Product. The traditional orientation of quality control has been to conduct extensive inspection activities in order to catch defects. Today's emphasis is on prevention rather than inspection. W. Edwards Deming, the American who is widely considered to be the father of the Japanese quality movement, said, "You don't get ahead by making products and separating the good from the bad, because that's wasteful."

A prevention orientation pays many positive dividends for the organization. Concern about defects pervades the organization, even impacting design and administrative activities. New products and services are developed in such a way that conformance to requirements in their manufacture and delivery is enhanced. Not only are failure costs reduced by a prevention mentality, but so are appraisal costs as well. With a greater degree of conformance ensured, inspection and other appraisal activities may be cut back with confidence.

Constant Improvement Is the Only Rational Goal. The contention that a certain level of defects is inherent or acceptable is widely rejected by quality professionals today as self-defeating. Traditional concepts such as "acceptable quality levels" (a certain level of defects is acceptable) and the "economies of quality" (beyond a certain point, improvements in quality are not cost-justified) are rapidly disappearing from the American workplace. If a certain level of defects is acceptable, then that level will always serve as a floor and will never be bettered.

Quality Is Everyone's Job. Accountability for quality should lie with those doing the work, not with the quality control department. The role of the quality professional is shifting from an enforcer to a facilitator—one who educates, trains, advises, and generally supports the organization's efforts to establish a culture for quality improvement. Just as productivity improvement must be an explicit responsibility of everyone in the organization, so must quality improvement be an integral part of everyone's job.

THE QUALITY PHILOSOPHY

Just as productivity philosophy or direction statements are useful in providing a vision and direction to the organization, so too can quality statements serve the same purpose.

A good example of a quality statement that also reinforces the productivity-quality connection is that of Hercules, Incorporated, a chemical products producer. The statement, which is signed by the chief executive A. F. Giacco and widely distributed, reads in part:

- ☐ It is the policy of Hercules Incorporated to provide products and services that are recognized by our customers as the standard of quality.
- ☐ Quality at Hercules means understanding our customers' requirements for the products and services we supply; setting a chain of specifications

that define these requirements; and then consistently conforming to the specifications in every step of every marketing, manufacturing, and purchasing activity.
☐ Productivity and quality are inseparable concepts. We must measure productivity as our ability to provide high-value products and services that meet customers' requirements at minimum cost.
☐ The priorities are safety, quality, and cost.

TOTAL QUALITY CONTROL

The new philosophies regarding quality form the underpinnings for the concept of total quality control, a state in which attention to quality pervades every aspect of organizational functioning.

A total quality control initiative generally starts with a definition of quality that incorporates in some fashion the idea that quality means meeting customer specifications. Westinghouse, for example, defines total quality as *performance leadership in meeting customer requirements by doing the right things right the first time.*

The definition is an important starting point, as all parts of the organization must be able to relate quality to specific definable standards or customer requirements. In the absence of such specifications, quality will be an imprecise and ill-defined objective.

A philosophy statement is often written by management to communicate both the definition and management's values with respect to quality. The relentless pursuit of quality, as defined, by all parts of the organization is clearly established by this statement as an organizational imperative.

In support of the total quality control concept, all organizational entities must identify their customers—whether internal or external—and define those customers' requirements. Improvement efforts then focus on meeting these requirements the first time every time.

Total quality control thus becomes the umbrella concept for a variety of improvement techniques. Statistical process control ensures that products conform to the specifications established by customers. Just-in-time systems force discipline into a manufacturing operation, so that problems are dealt with immediately as they occur. Quality circles are utilized to involve employees in identifying and solving quality problems. Cost of quality measures are calculated regularly to monitor progress. Diagnostic and opportunity assessment projects are undertaken to identify impediments to quality improvement. Multidisciplinary task forces deal with major, system-wide quality-related issues.

The entire organization reorients itself to the customer and his requirements. This may represent a significant change for technical and administrative groups, which often operate under internally-generated, rather than customer-defined, standards or specifications. By refocusing on the needs of the internal customer, the support organization provides services that are of greater value to its users, who in turn are better able to fulfill their missions.

QUALITY AS THE DRIVING FORCE

The productivity-quality connection has led many organizations to recognize quality as a major productivity improvement opportunity. This has been a particularly fortunate development, as improved quality pays other dividends as well—among them improved customer satisfaction and greater pride of workmanship.

A number of companies, in fact, have chosen to make quality the driving force for their performance improvement efforts. These organizations establish quality, rather than productivity, as the rallying point for organizational improvement. This decision is typically made in part because of the many benefits of improved quality; it is further motivated, however, by a simple fact: Quality is something that everyone in the organization can relate to positively. Unlike productivity, quality improvement does not carry negative connotations to the hourly work force and is not usually perceived as a threat to jobs.

An example of this approach is provided by Armco Corporation, a diversified steel manufacturer.[3] Armco utilizes a comprehensive improvement process called "Q + " (quality + productivity + participative involvement).

Developed in the early 1980s in response to difficult economic times, the Q + process was designed to involve employees at all levels in a comprehensive improvement process. Underlying the approach is a management philosophy that real and lasting improvement is a direct result of participative involvement.

The Q + process is managed at an operating unit by a middle management committee called the "Q + Team." This team is responsible for introducing the process to the organization, providing required training, and coordinating all program activities. The decision to charge middle management with the overall management of the Q + process was based on the recognition that the commitment and support of middle management was vital to the success of any change effort.

The improvement process consists of two phases, the first of which is designed to build a culture that is supportive of quality improvement. Extensive training is conducted during this phase, and participative management is introduced through the implementation of Corrective Action Teams (CAT) in various parts of the organization. A version of quality circles, CATs are designed to involve employees in identifying, and eliminating, causes of errors in the system. Other activities during this phase include awareness-building and the development of measurement systems.

The second phase of the Q + process is oriented toward system building and focuses on implementing systems and tools to bring about quality improvement. Systems installed during this phase may include computer-aided engineering, computer-aided manufacturing, statistical process control, and supplier involvement activities.

A corporate staff group exists to support the implementation of the Q + process in the operating units. Headed by a Vice President of Productivity, Quality, and Information Resource Management, the corporate organization provides initial training and guidance to those units that seek their assistance.

Armco's Q + process contains most of the features of an effective productivity management process—management commitment, awareness-building, measurement, employee involvement, and supporting organizational entities. The only difference is that quality, rather than productivity, is the driver for the process. The assumption is that productivity improvement will be an inevitable and predictable outcome of quality improvement.

QUALITY IN SERVICES

The notion of quality as the driver for a broad performance improvement effort is not limited in application to manufacturing companies. There are many examples of service organizations that have concluded that quality can serve effectively as the focus of their improvement efforts. As the mission statement of the General Services Division of Consumers Power Company, a Michigan utility, states, "Quality is the foundation of excellence."

One service organization that has made a major commitment to quality is The Paul Revere Life Insurance Companies, a subsidiary of Avco Corporation.[4] In 1983, executives of The Paul Revere Companies concluded that quality improvement would be the goal of a major organizational initiative. Like the Armco program described earlier, this quality effort contains many of the elements necessary to ensure that quality improvement is approached as an integrated management process.

The effort was structured and managed by an eight-member Quality Steering Committee composed of senior managers. The process, dubbed "Quality Has Value," had, as an avowed goal, a change in corporate culture.

Some of the techniques used to improve quality and to foster change included the following:

☐ Surveys of both employees and policyholders to identify areas that needed improvement.
☐ Quality Teams, a modified form of quality circles, to involve employees in the process. Quality Teams were non-voluntary, with all employees required to be on at least one team.

☐ Supportive communications and recognition programs, such as a "Quality News" newsletter and a formalized procedure to ensure that company executives meet periodically with Quality Team leaders.
☐ Value analysis workshops to provide managers with the tools to evaluate their departments' functioning and develop ideas for improving effectiveness.
☐ A family of measures, called a "quality index," to monitor departmental and company performance.

The Quality Has Value process has resulted in several million dollars of measurable improvements and is considered a success by Paul Revere's parent company.

Another service organization that has effectively driven its improvement effort through quality is Florida Power & Light Company.[5] Dramatic changes in the environment within which utilities must operate were the impetus for the effort, which was dubbed Quality Improvement Program (QIP).

The company adopted a definition of quality as "conformance to valid requirements" and organized the QIP effort around "Eight Steps to Quality":

1. Management Commitment. Initiated in the Chairman's office, the initiative seeks to build management support at all levels.
2. Quality Improvement Teams. A decision to pursue quality improvement through teamwork led to the creation of a hierarchy of teams that parallels the formal organization structure. At the top of the company, a fifteen-member Quality Council, made up of senior managers, sets policy and governs the effort. At the division and corporate department level are Lead Quality Improvement Teams, multi-functional groups of managers that deal with problems referred from lower-level teams as well as those that exist at their own level. At lower organizational levels, three types of teams exist: Local/Functional Teams, which solve problems in their own work areas; Cross-Functional Teams, which deal with broader issues; and Task Teams, which are constituted to solve specific problems. Finally, there are Corporate Issue Teams, groups of middle and senior managers who address company-wide issues.
3. Management Orientation and Training. Team leaders and facilitators are provided extensive training in the QIP philosophy, improvement processes, group dynamics, and consensus decision-making.
4. Economic Analysis. Cost of quality measures are widely used to identify improvement opportunities and to measure effectiveness. QIP teams initially focus on failure costs (redesigns, corrective actions, in-

valid procedures, billing errors, equipment failures, and service liabilities), but ultimately address prevention and appraisal costs as well.

5. Identify Root Causes. The various teams review the requirements of various products and processes, then identify deviations from those requirements.
6. Corrective Action. Utilizing standard problem-solving techniques, QIP teams develop permanent solutions to the root cause problems identified in the previous step.
7. Awareness of Fit Between Team Efforts and Corporate Goals. Program successes, and their impact on the company, are publicized through a variety of media, including a monthly video program and the company newspaper.
8. Recognition. The publicity and communications activities described above also serve to recognize and reinforce individuals, teams, and team members.

While the Florida Power & Light initiative is driven by quality, improved productivity is clearly an intended outcome; the company attributed $42 million in savings to the QIP effort in a recent year. The program succeeds because it too contains key elements of a management process—management commitment, awareness-building, supporting organizational entities, measurement, recognition, and employee involvement.

QUALITY IN THE PUBLIC SECTOR

The use of quality as the driving force for improvement is not limited to the private sector. One of the better examples of an effective quality improvement process is provided by the Naval Air Rework Facility (NARF) at North Island in San Diego.[6]

One of six rework facilities operated by the Navy, NARF North Island exists to maintain and overhaul Navy aircraft and components. It also has some manufacturing operations that produce certain types of equipment. It is a sizeable organization, with 5,200 employees and an annual payroll exceeding $148 million. Its civilian employees are organized by seven labor unions, the largest of which is the International Association of Machinists and Aerospace Workers.

Like many organizations, NARF North Island had utilized a number of improvement techniques, including quality circles and suggestion systems. These efforts evolved into a Total Quality Management Program (TQMP) which had, as a major objective, a change in organizational culture to one in which quality is the first priority and an integral element of everyone's job.

The TQMP is customer-focused; one of its key underpinnings is the identification of customer-supplier relationships and a focus on the customer's needs as the criterion by which quality is defined.

The process undertaken at NARF North Island contained most of the key elements of a productivity management process, such as awareness-building, organizational infrastructure, measurement, and employee involvement.

Particularly noteworthy was the use of multilevel structures involving substantial numbers of employees in the improvement process. This structure is presented graphically in Figure 11-1.

Managing the Total Quality Management Process is a five-person steering committee composed of top managers. The functions of the steering committee are typical and include development of training efforts and provision of long-term management commitment to sustain the process.

Quality Management Boards (QMB) exist at multiple levels—division, branch, section, and shop—and provide the critical infrastructure to carry out the quality improvement process. The boards cut across organizational lines, including representatives from different departments at the appropriate level. Each board also includes a representative from the next higher and next lower level of authority, thus providing hierarchical continuity. Issues

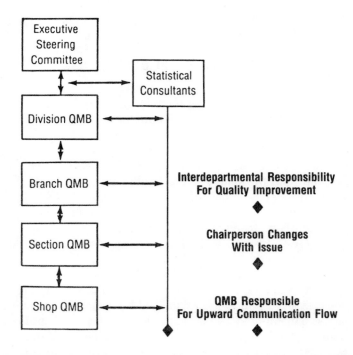

Figure 11-1. Quality management boards produce two-way communication flow for process improvement.

that cannot be resolved at the level where introduced may be referred to either the next higher or next lower level for resolution.

The QMB at each level identifies the process that occurs at its level, defines the requirements for quality in these processes, and evaluates the causes and effects of quality problems. Consensus is reached regarding specific causes to be addressed, and Process Action Teams are created by the board to eliminate the problem.

The leader for each Process Action Team is appointed by the steering committee, and a member of the steering committee is assigned to oversee the team's activities. The team leader selects the team members, who set about to develop a solution to the problem. Once a solution has been developed, the team establishes measurement devices to ensure that the situation can be monitored on an ongoing basis. The team is then dissolved.

The final element of the organizational infrastructure is a group of internal consultants who function as facilitators to the boards and teams. A major selection criterion for facilitators, who are chosen by the steering committee, is that they be respected by their peers. The facilitators are provided special training and perform their duties on a part-time basis, retaining their regular positions within the NARF.

The improvement effort at NARF North Island is driven by quality but displays the principles of an effective productivity management process, particularly in its use of extensive organizational structures and employee involvement at all levels.

THE QUALITY CULTURE

An organization that chooses to pursue quality as the path to productivity should be cognizant of the traps that befall many well-intended productivity improvement efforts. Just as a productivity "program" has little chance of fostering lasting improvement, so any superficial attempt to improve quality will fail to ultimately boost productivity.

Long-term success can only be achieved through changing the organizational culture and institutionalizing quality as an operating norm of the organization. The principles for achieving a productivity culture have equal application and are easily transferable to quality or any other organizational performance variable.

REFERENCES

1. Fuller, F. Timothy, "Eliminating Complexity from Work: Improving Productivity by Enhancing Quality," *National Productivity Review*, Autumn, 1985.

2. Crosby, Philip B., *Quality is Free: The Art of Making Quality Certain*. New York: McGraw-Hill, 1979.
3. *Case Study 33: National Production Systems—Los Nietos*. Houston, TX: American Productivity Center, 1984.
4. *Case Study 42: The Paul Revere Life Insurance Companies*. Houston, TX: American Productivity Center, 1984.
5. *Case Study 39: Florida Power & Light Co.* Houston, TX: American Productivity Center, 1984.
6. *Case Study 53: Naval Air Rework Facility, North Island, San Diego*. Houston, TX: American Productivity Center, 1986.

12

Knowledge-Worker Productivity

THE NATURE OF KNOWLEDGE WORK

Organizations that desire to develop a productivity management process invariably have a problem applying the concept of productivity to their knowledge workers. Somehow the traditional improvement techniques, which were developed in the factory, don't seem to fit.

It is important at this point to make a distinction between knowledge workers and white collar workers. Knowledge workers are white collar workers, but not all white collar workers are knowledge workers. Many white collar workers are involved in producing clearly-defined output in a repetitive fashion. The activities of clerks encoding checks in the back office of a bank, for example, are at least as routine and repetitive as those of any blue collar worker in a manufacturing plant. Their outputs are countable, their productivity is measurable, and the improvement techniques utilized in the factory are largely applicable.

The difficulty arises, then, not with white collar workers, but with knowledge workers—those employees whose work is largely nonrepetitive, nonroutine, and discretionary. This category of employees includes scientists, engineers, financial analysts, attorneys, employee relations specialists, salesmen, and managers of all kinds.

There are distinct differences between knowledge work and clerical or factory work. Some characteristics of knowledge work are:

- ☐ **Intangible outputs.** Unlike the factory worker producing a product, the output of the knowledge worker is invariably a service and is usually ill-defined, intangible, and uncountable.
- ☐ **Nonlinear process.** The process for manufacturing a product is typically apparent and linear; certain physical activities are executed repetitively and sequentially. Theoretically, the process can be optimized. The process of the knowledge worker, however, is very different. The analytical and creative thought processes that are so important to knowledge work are virtually impossible to define, much less optimize, and the great variety

of situations typically encountered by knowledge workers make any effort to create a routine and linear process very difficult.

☐ **Unclear contribution.** Knowledge workers, particularly those in administrative functions, often lack a clear sense of how they contribute to the company's strategic objectives. Employee relations people, for example, may have difficulty relating to their company's objective to become the low-cost producer in its industry.

☐ **Unclear performance criteria.** The criteria for good performance are often not clearly defined in knowledge-worker organizations, and a variety of perceptions may exist regarding desired performance outcomes.

☐ **High interdependence.** Collaboration and teamwork are often vitally important to the success of knowledge worker organizations. Engineers from various disciplines must collaborate to effectively design a plant, and the synergy of group creative processes are often vital to achieving breakthroughs in a research department.

All of the concepts of a productivity management process, as described in previous chapters, certainly apply to knowledge workers. Management and employee awareness of productivity must be raised, responsibilities for improvement must be made explicit, measurement systems should be developed to monitor progress and to provide feedback, and employee involvement must be pursued to build employee commitment and to foster change. Successful knowledge-worker productivity efforts invariably employ the same broad strategies that have proven successful in manufacturing.

The distinctive nature of knowledge work does, however, call for some tactical modifications.

EFFECTIVENESS VERSUS EFFICIENCY

The very definition of productivity must be rethought in a knowledge-worker environment. While the notion of increasing the ratio of output to input works fine in a manufacturing (or even clerical) environment, the concept smacks of efficiency improvement to knowledge workers and is predictably rejected by them.

Their resistance to an efficiency-oriented improvement program is justified, for while efficiency is certainly important to organizational success, it should rarely have top priority in a knowledge-worker organization.

Consider, for example, a research and development department. Should employees in this organization seek to maximize the number of new products developed, or should they attempt to maximize the value, marketability, and timeliness of a limited number of new products? As another example, should a management information systems department attempt to maximize efficiency in the volume production of new programs, or should they concentrate on producing systems that function as intended and that meet their users' needs?

In a knowledge-worker environment, the productivity emphasis must shift from the efficient production of a product to the effective delivery of a service. Knowledge-worker organizations generally exist to provide a service to the larger organization of which they are a part, and their focus should be on delivering that service with maximum effectiveness. And what is effectiveness? Effectiveness can be defined as having maximum value to the larger organization in the pursuit of its mission and objectives. If the services provided by a knowledge worker department are not of significant value to the company in meeting its objectives, that department cannot be considered effective, no matter how efficiently those services were delivered.

A number of different performance variables may be key to improving effectiveness:

☐ **Quality.** Providing a service that is free of error and meets the user's needs is a key determinant of effectiveness. Nothing can compensate for a defectively-designed product, financial statements that are incorrect, a lawsuit that is poorly defended, or a computer program that does not function as intended.

☐ **Timeliness.** Very often the value of services diminishes greatly if they are not provided in a timely fashion. Late engineering work for a new facility that is needed to meet increased customer demand, for example, may cost the company millions of dollars in lost sales. Financial analyses that are not completed when needed may result in management decisions that are made without important data. And the failure to meet regulatory filing deadlines may prove very costly to the company.

☐ **Efficiency.** While normally not the dominant performance variable in knowledge work, efficiency is nonetheless an important element of effectiveness. Cost-effectiveness in the provision of quality services invariably enhances the value of those services.

☐ **Innovation.** In some knowledge-worker organizations, creativity and innovation are of foremost importance. Research organizations are the most obvious example, but creativity often is of great importance to the success of engineering, marketing, and legal departments as well.

The relative importance of these variables will, of course, vary from one knowledge-worker organization to another. Unfortunately, the relative importance of the various performance criteria is often not clearly defined, or management behavior is often inconsistent with the espoused priorities. Management may verbally promote quality of service, yet focus its attention on cost control. Such mixed signals confuse employees and impede effectiveness.

In any event, productivity, if not defined differently by management, will normally be equated by knowledge workers with efficiency and cost reduction. In a manufacturing organization that is attempting to establish an organization-wide productivity management process, this perception may well cause the company's knowledge workers to opt out of the process. In a company whose employees are predominantly white collar, the lack of buy-in to the effort may be widespread and ultimately fatal to the entire initiative. It is critical, therefore, that productivity be placed in the proper context for knowledge workers.

CLARITY OF MISSION AND SERVICES

If the management of a knowledge-worker organizational unit is to succeed in instilling an effectiveness-oriented management process, it is vital that employees have a clear sense of their unit's mission, how it relates to the company's mission and strategic objectives, and the nature of the services that they provide in support of their mission.

The concept of effectiveness only has meaning within the context of *services*. If effectiveness means providing services that are of value to the larger organization, those services must be clearly defined and understood by all. If there are varying perceptions regarding the reason for being of the unit and the nature of the services provided, effectiveness cannot even have meaning, much less be achieved.

Services are often confused with activities, a problem that is exacerbated by organizational boundaries and specialization of work. When employees of a management information systems department were asked to define the services they provided to their company, the systems analysts talked about designing state-of-the-art systems, the programmers talked about writing programs, and the operations people talked about running their computers with minimum downtime. Each group focused on its day-to-day *activities* and saw these as its reason for being. The result was that systems analysts designed elegant systems that were difficult to program and did not meet users' needs; programmers wrote programs that were not efficient; and operations people ran the computer room as they saw fit, without regard for user priorities or deadlines. The predictable result was that the department was viewed as ineffective by other parts of the organization, and department management was preoccupied with dealing with user complaints.

After gaining some insight into the nature of the problem, department management developed a mission statement that talked of meeting the company's information needs in a timely and cost-effective fashion. They also defined the services provided by the department; these included "developing and implementing software systems that meet users' needs at minimum cost" and "providing prompt and efficient execution of users' computer processing requests." Management then undertook to reorient department employees to these services and to create an improvement process through the use of the techniques and processes described in earlier chapters.

As employees became aware of the department's mission and services and began to problem-solve around them, some interesting changes occurred. The different groups began to recognize that their work activities were not end objectives, but were integral elements of a broader service. This recognition raised their appreciation of their contribution to the end product and of their impact on other groups within the department. Department members also became more user-oriented and began to view their activities in terms of the impact on their internal customer.

As a general principle, services should be defined in terms of the user's needs. Invariably, significant improvements in effectiveness by knowledge-worker groups are preceded by the creation of a customer orientation: identifying the internal customer, defining his needs, and reinforcing the organization for meeting those needs. Defining services that encompass multiple activities and bringing focus on the internal customer is a critical step in creating that orientation.

DEFINING AND MEASURING PERFORMANCE

As was indicated earlier, unclear performance criteria, or emphasis on inappropriate performance variables, is often an impediment to improving knowledge worker productivity. With an emphasis on efficiency or cost control, the organization cannot relate to a concept of service effectiveness that encompasses quality, timeliness, and other variables.

As was discussed in the chapter on measurement, quantitative indicators are invaluable to the reinforcement of a productivity management process. It was also noted in that chapter that a *family of measures* is generally appropriate in a white collar environment. It is worth repeating that productivity, if not defined otherwise, will be construed by employees as an efficiency concept.

The family of measures fits this environment because it reinforces the notion that performance is a multifaceted concept and cannot be encapsulated in a single measure. Services must be provided with quality *and* timeliness *and* efficiency. The weighting mechanism described in the family of measures discussion also serves the useful purpose of clarifying management priorities; if quality of service is to be the preeminent performance variable, its status as such can be made apparent through the weightings assigned.

In any event, the lack of clear performance criteria and indicators of effectiveness are major impediments to productivity improvement in a knowledge worker organization and must be addressed at an early stage in the process.

INVOLVEMENT AND TEAMWORK

While employee involvement represents the key process for obtaining improvement in any environment, there are some special considerations in a knowledge-worker environment.

Since knowledge workers often resist the idea of productivity improvement ("it doesn't apply to us") and measurement ("what we do can't be measured"), it is vital that their commitment to productivity measurement and improvement be obtained through involvement in these processes. All of the techniques described in Chapter 8 are applicable and can be found in knowledge-worker organizations. The Nominal Group Technique is particularly effective in obtaining employee input for a family of measures, and task forces and cross-functional problem-solving teams can be effective in defining customer requirements, assessing barriers to performance improvement, and solving interface problems.

Like involvement, teamwork should be an explicit outcome of any productivity management process. But teamwork may take on special significance in a knowledge-worker environment. The issues dealt with in many knowledge-worker groups are very complex, and their resolution requires the collaboration of people from different disciplines and with different perspectives. Productivity efforts in these organizations may require initiatives focused on team-building and interface issues.

THE APC IMPROVEMENT METHODOLOGY

The American Productivity Center in 1983 developed a productivity improvement methodology that was designed to meet the needs of white collar organizations. The impetus for this effort was the growing predominance of white collar employment in the United States and the generally unsatisfactory results of many industry efforts to improve white collar productivity.

In order to assess the effectiveness of this methodology, the Center organized a multiclient action research project to implement and evaluate the approach in a variety of white collar organizations. The 13 organizations that ultimately sponsored this project were Armco Corporation, Johnson & Johnson, Motorola, TRW, the National Aeronautics and Space Administration, Dun & Bradstreet, Warner-Lambert, Ortho Pharmaceutical, General Dynamics, Northern Telecom, Rockwell International, McDonnell Douglas, and Atlantic Richfield. Each of the sponsoring organizations supported six pilot projects in a variety of functional groups—engineering, finance, R&D, human resources, marketing, and information systems. A steering committee consisting of middle management representatives from the sponsoring companies was formed to oversee and evaluate the project.

A description of the six-step methodology (Figure 12-1), which is based on services and is oriented toward effectiveness, follows:

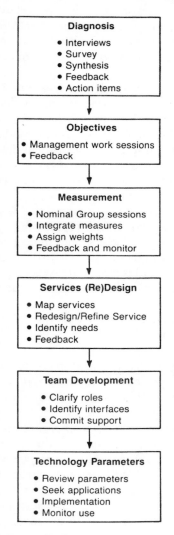

Diagnosis
- Interviews
- Survey
- Synthesis
- Feedback
- Action items

Objectives
- Management work sessions
- Feedback

Measurement
- Nominal Group sessions
- Integrate measures
- Assign weights
- Feedback and monitor

Services (Re)Design
- Map services
- Redesign/Refine Service
- Identify needs
- Feedback

Team Development
- Clarify roles
- Identify interfaces
- Commit support

Technology Parameters
- Review parameters
- Seek applications
- Implementation
- Monitor use

Figure 12-1. White collar improvement methodology.

Diagnosis. The first step in the methodology is designed to do two things: define the services that the subject organizational unit exists to provide and identify those characteristics of organizational functioning that are impediments to increasing the effectiveness with which those services are delivered.

Knowledge-worker organizations generally exist to provide services to the larger organizations of which they are a part, and, as was suggested earlier, lack of clarity regarding those services (or confusion between activities and services) represents a major inhibiter of effectiveness improvement. The purpose of the first phase of the improvement methodology, then, is to define and clarify those services that are provided by the department or unit. These services are defined broadly (rarely exceeding more than four or five in number) and often cross organizational lines. A service of a human resource department that relates to the provision of qualified and capable employees, for example, may encompass the efforts of the selection, training, and human resource planning sections. Since the methodology focuses on the effectiveness with which services are delivered, a clear definition and appreciation of the services provided are requisite for success.

The second outcome of the Diagnosis phase is the identification of barriers or impediments to increased effectiveness in the delivery of services. These barriers may encompass broad areas of organizational functioning: lack of appreciation by employees of organizational goals and strategies, which leaves employees without a clear sense of direction; poor communications, resulting in poor decisions and low employee commitment; nonreinforcing reward and recognition systems, which encourage nonproductive behavior; lack of employee input to problem solving and decision making; lack of service orientation and inadequate procedures to obtain feedback from internal customers; inadequate use of technology; and mistrust, low morale, and job insecurity.

The Diagnosis phase, like all the steps in the methodology, is executed in a participative fashion. Structured, confidential interviews are conducted with a cross section of employees representing different functional activities and organizational levels. In addition, a written survey is administered to all employees to ensure full participation and to provide statistical data to validate information obtained during the interview process. Conclusions or findings of this phase, and all subsequent phases, are fed back to the organization at large.

At this point, the improvement process proceeds on two parallel tracks. On one hand, action planning teams are formed to deal with the organizational issues and barriers identified. These teams include both management and nonmanagement employees and represent a cross section of the organization. Each team deals with a single issue and is guided by someone with the appropriate skills. The teams further clarify the issue being addressed and develop recommended steps to deal with it.

The second track pursues the implementation of the remaining phases of the methodology, with a focus on the services defined.

Objectives. The second phase of the process is designed to create a consensus among management regarding the type of culture that is required to support effectiveness improvement and to develop specific performance objectives for each of the services.

It is important that management have a common vision of organizational culture and management style. Do they believe, for example, that their organization should be characterized by open communications, employee involvement in decision making, and teamwork? Without commitment to a well-defined vision, management behaviors may be inconsistent and inappropriate to the effectiveness improvement initiative. Exercises such as the "Best/Worst Exercise," described in Chapter 7, are used to stimulate management thinking and to facilitate the development of consensus around organizational culture.

The second aspect of the Objectives phase is the development of specific performance objectives for each.of the services defined during the Diagnosis phase. By setting specific objectives around productivity, quality, timeliness, and other performance variables, management communicates its expectations and, in effect, defines performance in relation to the articulated services. The existence of clear objectives associated with a defined service serves to overcome the lack of clarity regarding the nature of effectiveness and to focus the organization on the achievement of specific outcomes. Objectives, in other words, provide direction and clarify management priorities.

The process used to establish objectives is a simplified brainstorming procedure: the management team members individually generate and nominate specific objectives, which are posted and discussed until consensus is reached on a small number (usually three to six) that, if achieved, will significantly increase organizational effectiveness in the delivery of the defined services.

Measurement. Once services have been defined and objectives have been set, the next logical step in the progression is to develop measures that can serve to track the organization's progress in achieving the objectives. For each service, a *family of measures* (as described in Chapter 6) is developed.

The approach utilized in the measurement development process is the Nominal Group Technique (Chapter 8). A cross section of employees involved in the delivery of the service is brought together to brainstorm and rank ideas for measures. Members of the group are selected for their ability to contribute to the process, and different perspectives and organizational levels are represented. It is generally advisable to invite users of the service to participate as well, to ensure that their viewpoint is represented.

The service statement and objectives are posted and discussed with the group to ensure the proper focus, and the NGT facilitator provides some brief education on measurement in general. The outcome of the process is a long list of measures, rank-ordered by the participants to reflect their judgment regarding the various measures' appropriateness and usefulness as indicators, given the objectives established.

The final step in the Measurement phase is the selection by management of the handful of measures that, when considered in the aggregate, are deemed to represent the best overall indicators of effectiveness in the delivery of the service in question. Given the multidimensional aspect of effectiveness, it is important that management select its family of measures to ensure that all of the important performance variables—quality, productivity, timeliness, etc.—are included.

Management must not, of course, neglect to *use* the measurement system effectively. The family of measures has great potential value in focusing the organization on effectiveness improvement; to realize that value, the measures must have wide visibility and be utilized as a medium for feedback, problem-solving, and recognition.

The use of participation in the Measurement phase serves to minimize the resistance to measurement that is often encountered in knowledge-worker organizations and fosters ownership of the improvement process by employees.

Service Redesign. Now that clear direction has been set by the identification of services, the setting of objectives, and the development of measures, the organization is ready to evaluate *how* it delivers its services. It is now appropriate, in other words, to look at activities.

This is where many organizations *start* their improvement process. They utilize popular techniques to examine work flow, reduce paperwork, or cut costs. These efforts are handicapped, however, because the organization that has not clearly defined its services has no benchmark against which it can evaluate whether its activities are even appropriate. It must assume that the present activities are rational and that the principal opportunities for improvement lie in performing those activities more efficiently. This orientation is further reinforced by the lack of objectives for effectiveness improvement and the absence of a broad family of performance measures.

An evaluation of work activities only makes sense within the context of services and clear performance criteria. Many of the organization's activities may not contribute significantly to the delivery of more effective services; if this is so, should we attempt to perform these activities more efficiently, or eliminate them altogether? In addition, if we do not have agreement on what our services are and on what constitutes effective delivery of those services, many of the improvement opportunities will go unrecognized.

Like the other phases of the APC methodology, Service Redesign is executed in a participative fashion. Another employee team is constituted, and their task is to map out the present method of delivery of one of the defined services. The service map identifies inputs, major activities, decision points, interfaces with other organizational units, and outputs. The intent of the exercise is not to detail every single activity, but to overview the broad elements associated with service delivery.

With the service map before them, the team members then identify those areas or activities that represent bottlenecks, delays, sources of errors, or inefficiencies. These inhibiters to effectiveness might include inadequate or incorrect inputs, cumbersome procedures, lengthy or unnecessary approval requirements, inadequate customer feedback, poor internal information sharing, workflow bottlenecks, unclear specifications, and various detrimental management practices.

The employee team then evaluates and prioritizes the various opportunities, develops action plans to capitalize on those opportunities, and presents these plans to management for approval. Additional employee teams may later be constituted to deal with the other identified opportunities.

Team Development. Significant opportunities exist in most organizations to improve effectiveness through improved interfaces, and this issue is dealt with during the Team Development phase of the APC methodology. Poor lateral communications, conflicting objectives, and interpersonal problems often contribute to a lack of teamwork and substandard organizational performance.

Another cross-sectional team is utilized during this phase to identify actual and potential interface problems within the organization. Some of these problems may have initially surfaced during the Service Redesign phase, while others are recognized during this exercise. As in the Service Redesign phase, opportunities for improvement are prioritized and action plans are developed.

Team Development also serves another purpose—to clarify changes in roles and responsibilities that may have occurred as an outcome of the Service Redesign phase. Responsibilities, decision-making authorities, and even organizational structure may have been changed in order to achieve effectiveness improvement, and the Team Development phase represents an opportunity to clarify new roles and promote the changes.

Technology Parameters. The final phase in the APC methodology addresses the use of technology to further enhance the organization's effectiveness. This phase is placed at the end of the process in order to avoid a common pitfall: the automating of an ineffective or inefficient work process.

It is all too common for organizations to become enamored of office technology and to spend substantial amounts of money to automate their existing processes and procedures. However, if these processes are inefficient, cumbersome, or unnecessary, then the organization has simply succeeded in institutionalizing its ineffectiveness and ensuring that these processes will be even more difficult to change.

Does it not make sense, then, to improve the effectiveness of the existing work processes before automating them? Should we not make sure that we are doing the right things before we electronically enhance them?

In the execution of this phase, employee teams, with the assistance of information systems experts, identify opportunities for effectiveness improvement through the use of technology. Input from equipment vendors or consultants may be sought, and action plans are developed to capitalize on the most promising opportunities.

Findings and Observations

Insights gained during the two-year White Collar Productivity Improvement research project led the American Productivity Center to eight general observations.[1]

1. White collar productivity improvement is founded on basic issues of vision, orientation, and management practices. While white collar productivity efforts frequently focus on a specific improvement technique or methodology, more basic issues should be addressed prior to

the implementation of these techniques. For example, managers should possess the skills and inclination to facilitate open discussion, problem solving, and personal risk-taking on the part of their subordinates; an openness to change should be fostered within the function; an orientation to internal or external customers and the provision of high-quality services should be developed; and the function's vision and direction should be congruent with the parent organization's strategic direction.

2. Attention to "operational" issues will enable productivity improvements to take place. By viewing the provision of services as a business, analogous to product lines in a manufacturing organization, attention can be focused on inputs and outputs, procedures utilized in the delivery of services, and the effective management of resources.

3. Training and coaching are required to deliver services effectively. Major opportunities for increased effectiveness of service delivery often lie in educating users to make independent decisions and develop their own solutions to problems that fall within the purview of the support function. This may represent a major role change for personnel in these functions, as their orientation may be the more traditional one that places emphasis on harboring and controlling their areas of expertise and delivering solutions, rather than consulting and facilitating the development of user capabilities.

4. Organizational administrative systems and processes offer a major opportunity for productivity improvement. Many administrative systems that were originally adopted by organizations to ensure control and consistency were found to have become bureaucratic and cumbersome over time. Capital allocation request systems can often be streamlined to provide more expedient processing for smaller requests, for example, and personnel requisition systems may be modified to allow for faster approvals where required to capitalize on rapidly changing business conditions.

5. Measurement of white collar work is both possible and desirable. Measurement was proven to be successful in a variety of functional groups, particularly when measurement was approached as one among many tools, when measures were associated with services or outputs rather than with individuals, and when employees had significant involvement in the development of the measures.

6. Justification of technology is best linked to critical features of service development and delivery. Cost justification of office automation often proves difficult, and many of the greatest opportunities to further the business through technology lie in areas where benefits are difficult to clearly define and quantify. However, with services defined, direction

clearly established, and measures of effectiveness in place, applications of office technology generally become much more apparent.

7. Self-reliance is a key to ongoing productivity improvements. While outside consulting assistance is often useful in planning and initiating a productivity improvement effort, ongoing improvement is vitally important and can only be achieved through the development of effective internal processes. These processes may include multifunctional teams to pursue improvement opportunities, continuous training and coaching in improvement methodologies, and frequent feedback from users.

8. White collar productivity improvement is dependent on seven critical success factors:

☐ A climate supportive of change, innovation, and risk-taking
☐ A vision for the future of the function that is shared among all employees
☐ Emphasis on service issues and opportunities
☐ A flexible methodology, one the function can adapt to its own circumstances and business
☐ Leadership by the function's managers throughout the effort, not by the consultant or a lower-level employee
☐ Technology directly linked to service leverage points
☐ Input from and "buy-in" by most employees at all levels of the function

The value of the American Productivity Center's White Collar Productivity Improvement research project lies not so much in the testing of a particular methodology, but in the validation of certain principles that have universal application to knowledge-worker functions: a focus on services is required, user input is essential, measurement is feasible and valuable, the implementation of office technology should be tied to service objectives, and the involvement of employees at all levels is a prerequisite for success.

REFERENCES

1. *White Collar Productivity Improvement: Action Research Project Summary and Findings.* Houston, TX: American Productivity Center, 1985.

13

The Role of the Union

UNIONS AND PRODUCTIVITY

Organized companies that wish to develop a productivity management process face an additional consideration: what is the role of the union? This question must be seriously considered, for whether management likes it or not, the union influences many aspects of organizational functioning, not the least of which is productivity.

Unions often resist productivity improvement efforts for the same reasons that employees in general resist them: Their awareness of the implications of productivity is low, and they believe that productivity improvement results in loss of jobs. They may be less cognizant of the connection between productivity and real wages and may not fully appreciate the greater long-term threat to job security posed by low productivity. The obvious response to this awareness problem is to ensure that union officers are included, and perhaps even given special attention, in the company's awareness-building activities.

Beyond the awareness problem, however, there exists an additional complication with respect to the union. The union is a political institution, with its officers elected by its constituency. Union leaders can only remain in office, and indeed the union itself can only continue to exist, if something of value is provided to the union's members. And the maintenance of an adversarial relationship is often viewed by union leaders as key to the members' perception of value provided.

If organizational improvement initiatives are pursued by management without involvement of the union, union resistance is the predictable result. If the union is not consulted or involved, it will likely view the improvement effort as a threat to the well-being of employees, or of the union, or both. The very fact that management does not seek the union's input and support is itself a red flag; surely, the reasoning goes, management does not have the best interests of its employees at heart when it does not consult with the employee's elected representative.

This phenomenon is not limited to hard, results-oriented productivity programs; many an employee involvement or quality of work life effort has foundered on the same shoals. The manager who says, "How can the union resist a quality of work life effort?" does not fully appreciate the political realities of unionism.

LABOR-MANAGEMENT COOPERATION

A noteworthy trend of recent years is the growth of collaborative labor-management improvement projects. In this era of world competition, both management and union leaders in many companies have reached the conclusion that the traditional relationship between these parties is an impediment to meeting today's competitive challenges.

Historically, two structures or processes have existed in unionized organizations to solve problems and deal with organizational issues (Figure 13-1). The first of these is the collective bargaining process, which deals with certain specific organizational issues: wages, benefits, working conditions, and the like. The collective bargaining process is adversarial by design and is accordingly a win-lose proposition. On every issue dealt with in this process, one side wins and one side loses. It is based on the principles of negotiation: in order to win on an important issue, we are willing to sustain a loss on another issue.

The second organizational process that exists to solve problems and deal with organizational issues is management authority. Management prerogative is brought to bear on a wide variety of issues: product decisions, financial matters, production scheduling, hiring and firing, and organizational structure, to name a few. In fact, management authority typically is the pro-

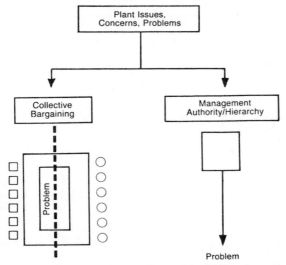

Figure 13-1. Traditional problem-solving processes.

cess used to deal with virtually all issues that are not subject to the collective bargaining process.

While these two processes have served us well for many years, both management and union leaders in many companies have reached the conclusion that these two processes are not adequate in times of turbulent change and intense competition. Many unionized organizations, accordingly, have sought in recent years to establish a third process, founded on collaborative problem-solving principles (Figure 13-2).

This collaborative process exists to deal only with those issues that management and union agree are of common interest. The process does not deal with contractual issues, where management and labor may have conflicting objectives, nor does it replace management authority. Collaboration is possible only where both parties perceive that common interests can be served.

Collaborative structures are actually quite common relative to one particular issue: safety. Many companies presently have labor-management safety committees that are designed to address this concern that is clearly of mutual interest. While these committees are not always models of collaboration, at least a joint structure does exist.

What is new in recent years is the considerable expansion of collaborative labor-management processes beyond the safety issue. It is not uncommon to find joint labor-management structures, for example, managing employee involvement or quality of work life improvement efforts.

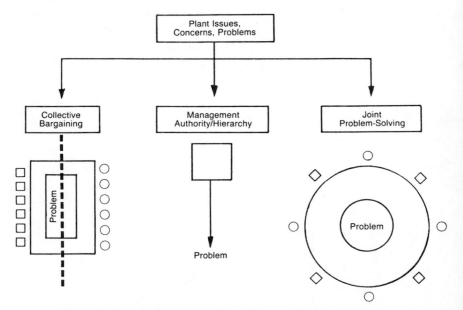

Figure 13-2. Three-pronged problem-solving processes.

Major collaborative efforts are most commonly found in troubled industries, where the mutual objective is a basic one: survival. It is here, where both parties finally recognize that their unwillingness to cooperate may be fatal to all involved, that the final barriers to collaboration can most easily be overcome. It is in these projects where we find union and management working together toward explicit business outcomes, such as productivity and cost reduction. As one might expect, joint labor-management projects are common in the basic industries, such as steel and aluminum, where some companies have invested considerable time and resources in support of these structures. Bethlehem Steel, for example, has supported company-wide labor-management processes for over six years.

Jointly-managed, collaborative labor-management projects can contribute greatly to the successful development of a productivity management process by building union support and commitment. If the process is truly *joint*, the union will have a degree of ownership for the effort that could probably not be achieved otherwise.

DEVELOPING A JOINT PROCESS

As enticing as a joint labor-management improvement process may sound, its implementation is fraught with difficulties. First of all, one of the parties simply may not wish to be involved in a joint process. Low trust, lack of a felt need, and political considerations may all contribute to this position.

Even if there is a desire to establish a joint process to work toward common objectives, the effort carries a high risk of failure due to the history of the labor-management relationship in most organizations. In too many companies, unfortunately, the adversarial nature of the relationship between union and management is just too ingrained to be overcome easily, even when there is a sincere desire by all parties to do so. While the process may get off to a good start, at the first point of major disagreement there is invariably a tendency to revert back to the longstanding adversarial mode of behavior. The parties start negotiating the key points of disagreement, and the collaborative environment is destroyed.

For this reason, most successful labor-management cooperation efforts have made use of a neutral third-party facilitator. This person usually guides management and union leaders through a multistep process such as the following:

1. Exploration. The facilitator meets separately with key management and union personnel. The purpose of these sessions is twofold: to educate the attendees on the nature and techniques of labor-management cooperation, and to provide a forum where the parties can separately

raise and explore their concerns about the process. A hoped-for outcome of this step is a commitment from both parties to proceed with a jointly-sponsored effort.

2. **Labor-Management Offsite.** Key people from both groups—usually the top-ranking leaders, but often people selected from lower-level positions as well—are selected to join the facilitator in an intensive, 2–3 day session held away from the organization's facilities. The facilitator's role here is key, as he must forge a climate of trust and an environment of collaboration where none has existed before. He utilizes a variety of tools, such as team-building activities and experiential exercises, to achieve these ends. It is here that the joint objectives of labor and management are identified and the ground rules for the joint process are established. The two groups often collaborate on the writing of a philosophy statement that defines their intentions and will guide their actions as they pursue the joint objectives. Finally, a steering committee is formed from among the participants to govern the ongoing process. The joint nature of the effort is generally reflected in the make-up of the steering committee: 50% labor and 50% management. If the facilitator has done his job, the parties leave the offsite meeting with a new-found commitment to work together in the pursuit of whatever objectives have been selected as the targets for the process.

3. **Implementation of Improvement Process.** Under the steering committee's guidance, a systematic improvement process is planned and executed. The steps in the process are usually similar to those employed in employee involvement efforts: Awareness programs are developed to educate the organization on the rationale and objectives of the joint process, a readiness assessment is conducted to identify the barriers to successful implementation, pilot sites are selected, and labor-management problem-solving teams are constituted to find ways to bring about improvements in the designated objectives areas. The steering committee continuously monitors and evaluates progress, fine-tuning the process and expanding it throughout the organization.

OTHER APPROACHES TO UNION INVOLVEMENT

Not every organization, of course, is ready for a jointly-sponsored, labor-management productivity improvement effort; in many cases, the relationship between management and the union is simply too negative, and the trust too low, to permit collaboration on anything. There also may not be a felt need on the part of one or both parties to work together. If the organization is not facing a clear and imminent threat to its survival, management and the union may be quite happy to maintain the existing relationship as it is.

The impracticality of a joint process does not, however, mean that the union should be ignored. Leaving the union totally out of the effort will greatly increase the likelihood of their resistance and may well contribute to the ultimate failure of the initiative.

At a minimum, the union should be kept informed. Special attention should be directed toward building the union's productivity awareness, as was suggested earlier. Management's intentions and objectives for the effort should be clearly communicated to the union, and assurances should be provided, if possible, regarding the security of jobs. Regular updates on the progress of the effort, and advance notice of major program changes, should be provided as well.

Greater levels of union commitment, of course, can be obtained through deeper involvement in the improvement process. Soliciting union input to assessment and planning activities, for example, should serve to raise union ownership and reduce resistance. And while a true joint process may not be feasible, companies often invite some degree of union participation in productivity or employee involvement steering committees. Alabama Power Company, for example, established in 1986 a productivity steering committee whose members included several corporate vice presidents and the president of the company's union. Such a commitment to union involvement surely pays dividends in terms of a more widely supported and effective productivity effort.

SAVING JOBS THROUGH COOPERATION

A clear example of the use of labor-management cooperation to improve organizational performance and thus save jobs is provided by an experience at Xerox Corporation.[1]

A study conducted by Xerox in 1981 indicated that some of the company's products were not competitive in the international marketplace. It was determined that a savings of $3.2 million could be achieved by subcontracting certain component parts that were presently manufactured in-house. The savings, which were considered vital to re-establishing the company's competitive position, would be realized by closing down an entire department and laying off 180 employees.

In an effort to avert this major loss of jobs, the Amalgamated Clothing and Textile Workers Union requested that management establish a joint union-management study team to attempt to find ways to improve efficiency in the affected department. Initially reluctant because of their belief that only labor cost reductions could return this department to a competitive position, management nonetheless agreed to try the cooperative strategy.

The study team, which consisted of six hourly and two management employees, was given six months to develop its recommendations. Team members received some training and pursued their task on a full-time basis. They were given the freedom to explore any activities that might result in cost reductions.

To support this effort, an Executive Labor-Management Policy Committee was formed. This group consisted of top union and company executives and had ultimate responsibility for approving the study team's recommendations.

Through a variety of means—soliciting suggestions from employees, conducting informal discussions, and visiting other companies—the study team identified 40 possible improvement projects. Ultimately, nine were selected for in-depth investigation:

1. Improved production equipment
2. Changes in work flow
3. Changes in work responsibility
4. Methods to reduce scrap
5. Improved work-order reporting procedures
6. Improved computer usage
7. Stabilization of the employee population
8. Production control and overhead adjustments
9. Reducing occupancy costs, such as floor space and utilities

At the conclusion of the six-month period, the study team presented proposals that would result in a total of $3.7 million of cost savings. The proposed changes were significant, including redesigning the physical layout of the department, expanding employee responsibilities, reducing overhead, and creating self-managing work groups.

Many of the changes, which were implemented over an eight-month period by the Executive Labor-Management Policy Committee, flew in the face of traditional union and management prerogatives. They probably would not have been possible without the collaborative spirit that resulted from the creation of a joint labor-management process.

A discussion of Xerox's company-wide effort is presented in Chapter 15.

COOPERATION IN THE PUBLIC SECTOR

The benefits of labor-management cooperation are not limited to the private sector; one need only review the experience of New York City's Bureau of Motor Equipment (BME) as evidence.[2]

A division of the City's Department of Sanitation, the BME is responsible for the maintenance of a 500-vehicle fleet, including collection trucks, mechanical sweepers, and salt spreaders. Its work is conducted by its 1,200 employees at one major central repair facility and a network of 73 repair garages located throughout New York's five boroughs.

Unfortunately, the BME was beset by problems in the late 1970s. An examination by the New York State Financial Control Board in 1978 found that the BME was capable of supplying the Sanitation Department with barely half of the entire fleet on an average day. As a result the Department was unable to perform its required tasks during normal hours and was forced to spend $5.5 million annually for overtime associated with night work. An additional $3.5 million in overtime was incurred by BME in performing its repair work. Contributing to this poor performance was an adversarial labor-management relationship and low employee morale.

A newly-appointed Deputy Commissioner of the Sanitation Department, Ronald Contino, addressed these problems with a labor-management cooperation strategy in 1979. Contino established a top-level Labor Committee and solicited recommendations for candidates from the various trade unions. Committee members would be relieved of their regular duties in order to work on the committee full-time and would report directly to Contino. The committee's charge was to work on operational and working condition problems, and it would meet weekly with Contino and his managers. Committee members were free to inform the union leadership of any matters discussed or any items that came to the committee's attention.

The committee visited field locations regularly to solicit ideas from employees. Suggestions received initially addressed working conditions, but after management proved its commitment by quickly addressing these issues, the focus of employee suggestions shifted to cost and productivity issues. Employees began to view the committee as a vehicle for gaining some control over their jobs and as a pipeline to the top of the organization.

Within a year and a half, the Bureau's problems had largely been solved, and Contino moved to institutionalize labor-management cooperation by establishing labor-management committees throughout the organization. Within a year, sixteen committees had been established, eight covering field repair operations and eight at the central repair facility. The membership of these committees consists of union officers, trades people, and middle managers.

The results of these efforts are impressive. The percentage of trucks available for work increased from 53.3 percent to 85.6 percent while substantial cost savings were realized. Productivity in the Central Repair Shop increased by 24 percent during the first year of existence of the eight committees at that location. Gains were also realized by the field operations committees, but actual savings were not calculated due to difficulties with the field information systems.

LABOR-MANAGEMENT SUCCESSES

The Xerox and Sanitation Department success stories are by no means isolated experiences. The U.S. Department of Labor has reported the results of over 200 successful labor-management efforts in a wide variety of industries.[3] Among them are the following:

☐ A possible plant closing was averted through a cooperative effort between Alcoa and the United Steelworkers.
☐ A Beech Aircraft suggestion program (Chapter 8) administered by a labor-management Productivity Council has been credited with saving several million dollars over a period of years.
☐ The Bethlehem Steel/United Steelworkers cooperative process involved 8,900 employees in labor-management participation teams throughout the company and resulted in widespread improvements in product quality, production costs, scrap, equipment downtime, and a variety of other areas.
☐ The Chicago, Milwaukee, St. Paul and Pacific Railroad realized over $3 million in savings through a cooperative process involving a number of unions.
☐ The Fiber Products Division of Diamond International Corporation reported a 16% improvement in productivity, a 40% reduction in quality problems, and a 55% reduction in grievances through its cooperative project with the United Paperworkers union.
☐ A joint labor-management program involving Malden Mills and the International Ladies Garment Workers Union resulted in $1.5 million in savings in a single year.
☐ Utilizing a joint effort to redesign work and increase self-regulation, Rohm and Haas Tennessee, Inc. and the Aluminum, Brick and Glass Workers Union achieved a 50% improvement in productivity.
☐ A labor-management committee established by the Plastics Division of Uniroyal and the United Rubber Workers developed a package of changes in work rules that resulted in an estimated $5 million in cost savings.

This small sampling of labor-management success stories demonstrates that cooperative efforts between union and management offer significant performance improvement opportunities.

REFERENCES

1. Lazes, Peter and Costanza, Tony, *Labor-Management Cooperation Brief: Xerox Cuts Costs Without Layoffs Through Union-Management Collaboration.* Washington, DC: U.S. Department of Labor, Bureau of Labor-Management Relations and Cooperative Programs, 1984.
2. *Productivity Brief 15: Labor/Management Cooperation Steers a Course to the Bottom Line.* Houston, TX: American Productivity Center, 1982.
3. *Resource Guide to Labor-Management Cooperation.* Washington, DC: U.S. Department of Labor, Bureau of Labor-Management Relations and Cooperative Programs, 1983.

IV
CONTINUOUS PRODUCTIVITY IMPROVEMENT

14

An Implementation Strategy

APPLYING THE PRINCIPLES OF CHANGE

The features and requirements of a productivity management process have been thoroughly discussed in the previous chapters. But how does one go about implementing this management process? The task is clearly great and requires that significant change take place in the organization. As with any major undertaking, we need a well thought-out, systematic approach if the effort is to succeed.

The fact that a productivity management process represents change is a key consideration in the development of our implementation strategy. Because this is a change process, we must carefully lay the groundwork and prepare the organization. We must identify the barriers to change and take steps to overcome them. We must be clear as to what we are trying to achieve, and we must effectively manage the change.

We must be careful not to immediately jump into improvement techniques, or we will fall into the Techniques Trap. While the techniques may succeed initially, they will not have a meaningful context to the organization, and there will be no processes in place to foster ongoing improvement. We will realize only one-shot improvements and will be doomed to a constant scramble to find and implement new techniques to replace those that have ceased to be of value. And organizational resistance will be predictable, as employees will fail to appreciate the implications of productivity and will view each new technique with skepticism and apathy.

This chapter will present an implementation strategy for a productivity management process. The strategy is based on the principles of change management and is tempered by the learnings gained by observing the productivity improvement efforts of many companies in a wide variety of industries.

The strategy is fairly general and generic, as it must be if it is to be of value to any and all organizations. As a result, it is important to view it only

179

as a general framework and to modify it as appropriate to fit organizational circumstances.

The strategy consists of six phases:

☐ Management Commitment
☐ Organizing for Change
☐ Assessment
☐ Planning
☐ Implementation
☐ Evaluation and Diffusion

Each of the phases will be discussed in turn.

Management Commitment

Any organizational change process, regardless of its nature or its objectives, must have the commitment of top management if it is to succeed. And productivity management certainly implies change. Attitudes and behaviors of employees, some of which have been shaped over many years, must change. Organizational systems, which may have been entrenched for decades, must change. And the organizational culture and norms must change.

Expecting changes of this magnitude and nature to take place without the full support and involvement of top management is simply expecting too much. Senior management is largely responsible for the organizational climate and tone, and they certainly control major organizational systems, such as the reward system.

The first step in the development of a productivity management process, then, is to obtain top management commitment to a long-term change effort. Without that commitment, there is little hope of achieving anything beyond a productivity program.

How does one obtain management commitment to a long-term change effort? Generally, this requires that management awareness be raised. Senior management is often too far removed from the day-to-day functioning of the organization to fully appreciate the magnitude of the task. All they need do, they believe, is to pronounce that productivity is a corporate objective. They assume that the message will work its way down, and that the organization will respond. They do not realize that existing organizational systems and processes do not effectively reinforce productivity improvement, and they do not appreciate the degree of resistance that will be encountered. They fail to recognize the need for a change effort and do not appreciate their role in promoting change.

In addition, senior management may have inadequate knowledge of the various elements and processes of productivity management. They may not fully understand the meaning and implications of participative management and they may not be aware of the many innovative approaches to reward systems. They may well view employee involvement and gain sharing as techniques or fads rather than as fundamental changes in management philosophy.

Finally, management may not clearly appreciate the relationship of productivity to the organization's business plan and strategic objectives. They may view productivity in a narrow, tactical sense rather than as a strategic issue that is critical to organizational survival.

The task, then, is to educate top management on the present status of productivity in the organization, the impact on productivity of organizational systems and culture, the role of the human resource in productivity improvement, and the principles of change. Only by creating a felt need to change can we expect to obtain the requisite commitment.

The productivity advocate would be well advised to bolster his management education process by seeking out and documenting the experiences of others. Nothing is more comforting to management than to know of the successes and failures of other organizations that have faced a similar challenge.

The management awareness-building process may also be helped by presenting the results of an organizational assessment exercise. While assessment is positioned in this strategy as the third phase, it is often useful to conduct at least a limited assessment as a first step in order to present management with some hard data that will raise concerns about the effectiveness with which the organization presently manages productivity.

As was discussed in Chapter 3, management commitment by itself is inadequate. An important part of the process is the demonstration of that commitment to the organization. But before commitment can be demonstrated, it must be obtained.

Organizing for Change

Once top management commitment has been obtained, the logical next step is to establish the responsibility for managing the process. This change process must be managed, and as was suggested in Chapter 4, this probably requires that organizational decisions be made. A highly-placed productivity executive may well be appropriate, as might a management steering committee. The various considerations surrounding these decisions were discussed in Chapter 4.

If management fails to provide for an organizational entity to manage the process, the probability of success drops dramatically.

It is advisable that management make and execute these organizational decisions immediately, as it is critical that those charged with the management of this effort have input into, and ownership of, the subsequent phases of the implementation strategy.

Assessment

Once management commitment is obtained and responsibilities have been established, it is tempting to begin planning the effort. The development of an implementation plan would be premature at this stage, however, as the present status of productivity in the organization has not been clearly analyzed and defined.

An effective cultural change process requires not only that the desired future state be defined, but also that the present state be clearly understood. The failure to conduct a thorough organizational assessment may result in some unpleasant surprises down the road. The plan may not effectively address some key organizational issues, or some unforeseen barriers may deal the effort a serious setback.

An organizational assessment (or diagnosis, as it is sometimes called) should have several objectives:

1. To explore all aspects of organizational functioning in order to identify the impact on productivity of present organizational systems, practices, and processes.
2. To identify the major barriers to an effective productivity management process.
3. To evaluate possible sites for a pilot effort.
4. To identify major opportunities for productivity improvement.
5. To promote change.

The last objective suggests that a good assessment can itself be an intervention; that is, it can be done in such a manner that it influences the very subject of the assessment—productivity, in this case. In other words, the assessment should be more than just a data-gathering device. It should be designed so that it serves to focus organizational attention on the productivity issue, thus raising the level of consciousness and awareness. It should also be designed to involve as many people as possible at all levels of the organization; by so doing, it models the principles of employee involvement. The assessment, then, should contribute more to the effort than its stated intent. It should also serve to raise awareness and begin to change behaviors.

There are a variety of ways to conduct the assessment. At a minimum, the productivity executive could convene a meeting of key individuals within the organization to discuss the various issues and reach consensus as to their status. While this approach may be the least costly and quickest to implement, it is also probably the least effective. Relying on the perception of a small group of people, particularly if they all occupy middle or senior management positions, will ensure that the conclusions reached are based upon a somewhat narrow and probably biased perspective.

A better way to conduct the assessment, and one that is consistent with the participative philosophy that drives the effort, is to obtain information from employees at all levels of the organization. We can then be confident that we have obtained a true picture of organizational attitudes, management behaviors, and the impact of organizational systems.

A broad-based assessment project usually employs one or more of the following data-gathering techniques:

☐ **Personal and confidential interviews with a cross section of employees.** This technique enables the assessment team to obtain detailed and extremely valuable information about the relevant issues. The interviewer can probe personal experiences, explore critical issues in depth, and obtain useful anecdotes. In addition, the confidential, one-on-one nature of the process ensures a greater degree of openness than might be obtained in a group setting. The drawback of the personal interview approach is that it is costly and time-consuming to interview a large number of employees.

☐ **Group interviews.** The major benefit of the group interview is that a larger number of employees can be involved in a cost and time effective manner. The trade-offs are that confidentiality concerns may inhibit some participants, and issues cannot generally be explored in as much detail as is possible in the individual interview.

☐ **Written surveys.** The administration of a written survey instrument is clearly the only cost-effective way to involve very large numbers of people in the assessment process. If input from an entire organization of several hundred or more people is desired, a survey is the only practical course of action. A survey offers another advantage as well: it provides quantifiable data to support (or contradict) the conclusions that have been reached through some other format. On the down side, a survey requires some interpretation and does not permit an in-depth evaluation of the issues.

☐ **Group brainstorming techniques.** The Nominal Group Technique (Chapter 8) is a useful assessment tool, for it provides a structured means of achieving a group consensus around a specific issue. The task statement, for example, might be, "What are the major barriers to productivity improvement in this organization?" The resulting rank-ordered list of barriers would be useful input to the assessment team.

It is generally advisable to employ a combination (if not all) of these techniques in the assessment process. Group interviews or NGT might be used, for example, to allow for the active participation of a sizeable number of employees and to identify the major organizational barriers and issues. More detailed information on these issues could be obtained through individual interviews, and a written survey would provide quantitative support for the conclusions reached.

The examination of company documents may also be a useful exercise in an assessment effort. Does the company's strategic business plan explicitly address productivity? Do budgets expressly provide for productivity improvements? Does the company's standard performance review form call for an evaluation of productivity performance? Do the monthly financial reports identify the impact of productivity changes on the bottom line?

It is often helpful to utilize external resources, either consultants or employees who are from another organizational unit, in an assessment project. Internal people may lack objectivity, and an internal interviewer will likely raise political and confidentiality concerns among the interviewees.

The main purpose of the assessment, of course, is to gather information about organizational functioning so that a rational and systematic plan can be developed. As such, the assessment should be designed to provide answers to many questions, including the following:

- ☐ What is the level of organizational awareness about productivity?
- ☐ What are employees' major concerns about productivity?
- ☐ How committed are supervisors and managers to productivity improvement?
- ☐ How explicit are responsibilities and accountabilities for productivity improvement?
- ☐ How effective are reward and recognition systems in reinforcing productivity?
- ☐ Do organizational goals for productivity improvement exist at various organizational levels?
- ☐ How widespread are productivity measures?
- ☐ How effectively is measurement utilized to reinforce improvement?
- ☐ Do employees receive feedback about organizational productivity and performance levels?
- ☐ Are communications effective in all directions?
- ☐ How effective have past productivity improvement efforts been?
- ☐ How involved are employees in performance improvement activities?
- ☐ How is employee involvement perceived by the organization?
- ☐ Do supervisors and managers have effective people-management skills?
- ☐ What is the union's position on productivity?
- ☐ What is the status of quality and how its relationship to productivity perceived?
- ☐ What is the predominant management style and how does it impact productivity?
- ☐ What are the major productivity improvement opportunities?

☐ What are the major barriers to employee involvement and productivity improvement?

Interestingly, perceptions about some of these issues often vary from one level to another. Supervisors, for example, may in general perceive themselves as having an involving style, while the perceptions of hourly employees are quite different. The fact that significant differences in perception exist is in itself useful information.

The assessment phase is not over with the gathering and evaluation of information. An important final step is to feed back the assessment team's conclusions to top management and to obtain their buy-in to the issues identified. Since continuing management commitment is so vital to this process, it is imperative that management agree with and support the conclusions of the assessment.

Planning

With useful assessment or diagnostic data in hand, it is now appropriate to develop a plan. A plan is an obvious need, given the nature of the undertaking. The successful implementation of a major new management process requires a systematic, orderly, and integrated approach; we cannot do it by the seat of our pants.

One of the first planning issues to be addressed is the advisability of a pilot project. In a large, multisite organization, it is often useful to proceed on a pilot basis in order to gain experience in the implementation process. The cost of a failure is less, and the experience gained will be invaluable in diffusing the process throughout the organization.

If a pilot effort is to be undertaken, the site selection decision is an important one; it is obviously desirable to pilot this effort in an area where the probability of success is high. Some of the selection criteria may include the following:

☐ Management willingness to support a long-term change effort
☐ Relatively people-oriented management style
☐ Satisfactory union-management relationships
☐ History of success in past endeavors
☐ Healthy organizational climate
☐ Absence of serious organizational or marketplace problems that would distract management's attention

Whether undertaken as a pilot or as an organization-wide effort, the plan development process should include people from a variety of functional areas and from multiple organizational levels. As with other elements of this process, the more people that can be involved, and the more constituencies that can be represented, the better the result.

Rather than have a single sequential list of action steps, the implementation plan should have multiple, simultaneous thrusts. Several avenues can and should be pursued simultaneously, thus shortening the implementation time while raising the level of visible activity in the organization. A typical plan might have four simultaneous tracks dealing with the following issues:

Track 1: Building Organizational Awareness

• Nature of the awareness message
• Identification of appropriate communications media
• Mechanics of dissemination
• Awareness of new employees
• Ongoing awareness and feedback processes

Track 2: Productivity Measurement

• Obtaining education in measurement principles and practices
• Determination of levels, functions, and inputs to be measured
• Integration of measurement into financial reporting systems
• Participation in the measurement process
• Provision of technical assistance to the organization in developing measures
• Visibility and use of measures in promoting productivity

Track 3: Employee Involvement

• Management awareness and commitment to participative management
• Union support for employee involvement
• Training requirements
• Pilot selection
• Facilitation resources
• Supporting systems and processes
• Diffusion plan

Track 4: Organizational Systems

• Explicit responsibilities and accountabilities
• Goal-setting
• Reward system modifications
• Recognition practices
• Integration into budgeting and planning systems

These four tracks are not necessarily the right ones for all organizations, and the specific details of the plan should, of course, be based on the assessment data and reflect organizational needs and circumstances.

While the plan will call for implementation along multiple tracks, it is important that these simultaneous activities be pursued in an integrated and mutually-reinforcing fashion. The awareness message should include the role of measurement and employee involvement in productivity improvement, for example, and the measurement development process should be executed in a manner consistent with the principles of employee involvement.

Implementation

With the plan completed and approved, the obvious next step is its implementation. Certain supporting activities and resources are important to ensure successful implementation:

☐ **Commitment-building.** The support of senior and middle management in any area affected by implementation activities is crucial to success. The typical organization has many initiatives or programs under way at any given time, and managers invariably must juggle priorities and manage a heavy workload. If steps are not taken to gain the commitment of management at all levels, lip service and half-hearted attention to the process is a likely outcome. This commitment is obtained by such activities as awareness-raising, demonstrations of commitment and support from higher organizational levels, and involvement in the implementation process itself.

☐ **Ongoing communications.** Implementation activities should be made highly visible through continuing communications to the organization at large. Maintaining this effort at the forefront of employees' attention serves to reinforce the process and increases the rate of change.

☐ **Facilitation resources.** Assistance in the form of advice, coaching, and trouble-shooting will be needed throughout the implementation process. A network of productivity coordinators with appropriate skills (Chapter 4) can provide invaluable support in this regard.

Evaluation and Diffusion

As implementation unfolds, ongoing evaluation and maintenance activities should be taking place. These activities include:

☐ Monitoring progress against plan benchmarks and objectives
☐ Evaluating the effectiveness of execution of the various plan steps
☐ Reviewing productivity measures to assess the degree of improvement
☐ Monitoring organizational climate and employee attitudes
☐ Recommending changes to the plan to reflect the experience gained or new developments
☐ Recognizing successes
☐ Documenting the implementation process and lessons learned
☐ Continuing to build management and employee support for productivity improvement

Flexibility is key in implementing a productivity management process. The plan must not be viewed as a rigid document, to be blindly executed without regard to successes, failures, or changes in circumstances. Through evaluation, the organization learns and adjusts accordingly. If results or circumstances demand a change in tactics or direction, that change should be forthcoming without hesitation or regret.

The likelihood of successful and useful evaluations is enhanced by the existence of clearly defined goals, objectives, and benchmarks, together with reliable measures of organizational productivity and performance.

The other element of this phase is the diffusion of the process throughout the organization. Many of the implementation activities will have taken place in organizational subunits or pilot areas, and the need to diffuse these processes and achieve broader organizational change will quickly become an issue.

Detailed plans for diffusion should probably be deferred until meaningful evaluation data are available, as the results of the initial efforts will significantly affect the nature and pace of the diffusion process. Obviously, those elements of the process that were most successful may be quickly diffused, while those that enjoyed only limited success, or outright failure, will be subject to re-examination, modification, and further piloting.

It should be noted that this implementation process is an iterative and flexible one; we cannot simply walk through the six phases once and be done with it. Implementing a productivity management process is as much art as it is science, and as was suggested earlier, flexibility is key. It is unrealistic, then, to expect to plan out the entire implementation process in detail and execute it without modification. We must be prepared to constantly cycle through the planning, implementation, and evaluation phases, learning as we go and modifying our plan accordingly.

The entire implementation strategy is graphically depicted in Figure 14-1.

ADVANTAGE OF PROCESS OVER PROGRAM

The failure to develop a systematic change strategy such as the one described in this chapter is symptomatic of the "program" approach. Those organizations that address productivity as a program rather than as a management process essentially enter the described process at the Implementation stage. They do not build top management commitment to a long-term change effort, so the necessary cultural changes never take hold. They may not create an appropriate organizational infrastructure to manage the effort, so integration and coordination are lacking. They do not conduct a rigorous assessment, so the many barriers to this effort are not fully recognized and become major impediments. While they may do some planning, that

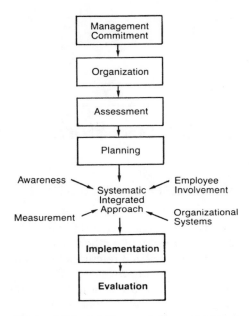

Figure 14-1. Implementation strategy.

planning is normally tactical in nature, oriented toward identifying and implementing improvement techniques rather than promoting change.

So the organization implements various improvement techniques and has itself a productivity program. The techniques provide some one-shot or short-term improvement, but nothing really changes. The processes required for continuing, comprehensive improvement never become institutionalized.

The organization that approaches the productivity issue as a change process, on the other hand, will likely enjoy long-term success. Management will be aware of the vital role they play in establishing a climate that is conducive to productivity improvement, and steps will be taken to prepare the organization for change. The stumbling blocks that exist will be identified and overcome early in the effort, and organizational systems and practices will be modified to reinforce productivity improvement as an ongoing process that capitalizes on the capabilities and commitment of people at all levels.

Short-term gains will still be realized, in part because they will be planned for, but also because a well-executed change process will begin to have an immediate impact on organizational attitudes and resulting behaviors. But more importantly, a new organizational forcus will emerge that will foster continuous improvement as an integral element of organizational functioning.

15

Success Stories

VARIETY IN APPLICATION

No two productivity management efforts look alike. Some consolidate and build on existing activities, while others essentially start from scratch. Some provide substantial centralized resources to support line efforts, while others do not. Some drive their improvement efforts through quality, while others focus on productivity measurement as their centerpiece. Most of the best ones, interestingly, seem to evolve over time into something that looks rather different from the initial incarnation.

The fact that successes can be found in a variety of approaches reinforces the earlier suggestion that a productivity management effort be customized to the organization's circumstances. It also suggests that there is more than one way to skin a cat.

It is instructive to review some of the successful efforts to observe the variety of approaches and the direction of their evolution. The following vignettes are not offered as detailed case studies, but serve as brief overviews of some noteworthy efforts.

TENNECO: FOCUS ON ASSESSMENT

Tenneco is a diversified, $15 billion company headquartered in Houston, Texas. About half of its sales come from energy-related businesses—natural gas transmission and oil production, refining, and marketing. The other half is accounted for by a variety of businesses: the Newport News Shipbuilding and Dry Dock Company, Tenneco Automotive (Monroe shock absorbers and Walker mufflers), J.I. Case farm equipment, the Packaging Corporation of America (PCA), and various other operations.

The genesis of Tenneco's productivity effort was a speech made at a company management meeting in 1978 by J. L. Ketelsen, then President and now Chairman and Chief Executive. In his address, Ketelsen called upon his managers to maintain Tenneco's history of success by effectively managing the company's assets and providing an environment in which people can contribute to the maximum of their capabilities.

In attendance at that meeting was Max Zent, then Director of Management Education for Tenneco. In response to Ketelsen's challenge, Zent organized a management meeting on productivity and quality of work life, the first of its kind at Tenneco. Encouraged by the attendees' highly favorable responses, Zent proposed to experiment with a productivity improvement effort in a single plant. The proposal was approved by Ketelsen, and Zent was placed on special assignment in late 1979.

Zent sought the aid of the American Productivity Center, enlisting the assistance of Center consultants in implementing the Center's newly-developed assessment process in Tenneco's Walker Muffler facility in Aberdeen, Mississippi. Dubbed "Productivity Focus" by the Center, the assessment process was later given a second test at Walker's plant in Greenville, Texas.

Productivity Focus was a process for evaluating those areas of organizational functioning that affect productivity and quality of work life.[1] As an organizing framework, these areas of organizational functioning were summarized into seven broad categories and presented graphically as shown in Figure 15-1.

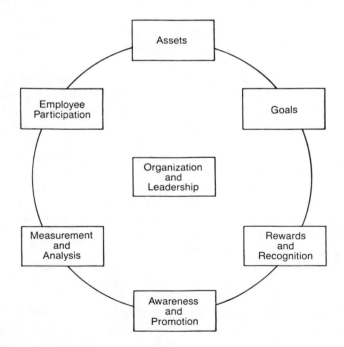

Figure 15-1. Productivity focus—areas of organizational functioning.

A variety of issues could be explored through the Focus framework. *Goals* include all of those organizational processes through which objectives are set and expectations established. Issues explored in this element include the clarity of organizational direction and strategic objectives, the existence of productivity goals, and the effectiveness of organizational processes utilized to establish goals. *Rewards and Recognition* encompasses all of those activities and systems through which employees are reinforced, both financially and nonfinancially, for improving productivity and performance. *Communications* addresses information-sharing practices, productivity awareness-building efforts, and the effectiveness of vehicles for upward communications. *Measurement* is concerned with the availability and reinforcing use of productivity measures as well as other quantifiable indicators of organizational performance. *Employee Involvement* addresses both the formal structures and informal practices through which employees have the opportunity to bring about improvements and influence decisions. Management attitudes toward employees and the organizational climate are also explored in this element. *Assets* is a broad category, encompassing all of those opportunities that may exist to improve the effectiveness with which the organization's assets—people, materials, equipment, energy, and facilities—are utilized. Issues often surfacing here include training, equipment utilization, scheduling practices, and scrap losses. *Leadership and Organization* occupy the hub of the Focus wheel, as this element drives all of the others. Management practices, as well as issues of organization structure, are addressed here.

The Productivity Focus assessment process was designed by the Center to be highly participative. Information was gathered by the project team through personal interviews with a cross section of employees at all levels of the organization. The data gathered through the interviews were analyzed and synthesized by the project team to produce key findings, which were fed back through a structured meeting with the top management team of the subject organization. Action plans were then developed to capitalize on the key opportunities, as determined by management. The data gathering process was subsequently enhanced at Tenneco through the use of the Nominal Group Technique (Chapter 8) to broaden participation and obtain preliminary data with which to better focus the interviews.

During 1981 and 1982, the thrust was on institutionalizing the Productivity Focus assessment process at Tenneco. Projects were conducted in other Tenneco units, including Monroe and Packaging Corporation of America (PCA). Permanent staff members were recruited to provide a cadre of internal resources to conduct assessments throughout the company. The first person to join Tenneco's productivity staff, an accountant, reflected Zent's conviction that measurement was a critical element of the improvement process. The second staff acquisition was an industrial engineer.

In 1983, after conducting Productivity Focus assessments at five Packaging Corporation plants, Zent felt the need to obtain greater leverage for his effort. The Packaging Corporation alone had 50 plants, and a plant-by-plant effort throughout Tenneco's many subsidiaries would be an enormously time-consuming and unwieldy process.

Having conducted a number of assessment projects by now, Zent observed an interesting phenomenon that provided the answer to his problem. It was apparent that certain issues were recurring; they appeared at plant after plant and thus were predictable and probably indicative of higher-level problems. By developing pervasive improvement processes around these issues and obtaining commitment to these processes at higher organizational levels, the needed leverage could be obtained.

The first such issue that Zent chose to attack was quality. Through the use of a cost of quality analysis (Chapter 11), Zent convinced the president of the Packaging Corporation that quality represented a major improvement opportunity at all PCA plants, and a comprehensive quality improvement process was launched. The process included extensive training, the formation of steering committees at all plants, and the use of problem-solving teams to identify and solve quality-related problems.

In 1984, the J.I. Case subsidiary became the focus of Tenneco's improvement activities, with the emphasis again on major issues that represented common opportunities at all locations. Initial assessment activities at Case led to an improvement process based on four major themes: quality, materials management, organizational structure, and measurement. Results were dramatic: $10 million in savings were realized over a two-year period at a single plant alone.

In view of the clear and pervasive opportunities represented by quality improvement, professionals skilled in the Deming, Juran, and Crosby quality processes were added to Zent's staff during 1984.

Tenneco's improvement process has thus evolved over the years into something that, in some ways, bears little resemblance to its origins. Assessment is still a key element of the process, but the bottom-up approach, which typically generated a large number of individual improvement opportunities, has been replaced by one having greater emphasis on data-gathering at the management level to identify major pervasive issues which then become the focus of improvement activities.

The present process is depicted in Figure 15-2. The overall effort in a given unit is managed by a management steering committee, whose commitment to the process is obtained by Zent and his staff in a structured, off-site meeting. Key issues are identified through management interviews, and the scope of the effort is defined.

Employee involvement in the process comes about through the creation of ad hoc teams established at each site to identify specific causes of poor per-

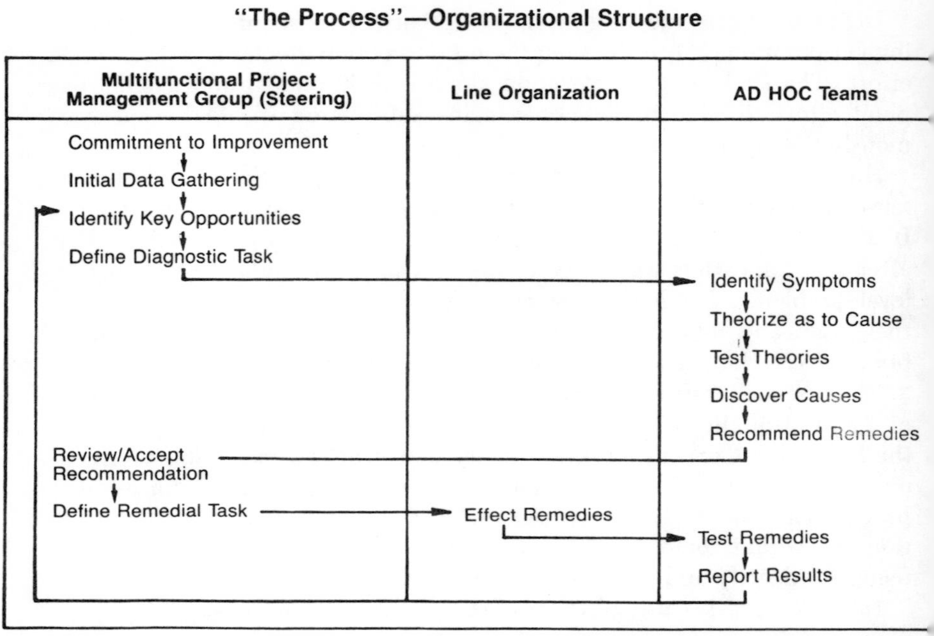

"The Process"—Organizational Structure

Multifunctional Project Management Group (Steering)	Line Organization	AD HOC Teams
Commitment to Improvement ↓ Initial Data Gathering ↓ Identify Key Opportunities ↓ Define Diagnostic Task		
		Identify Symptoms ↓ Theorize as to Cause ↓ Test Theories ↓ Discover Causes ↓ Recommend Remedies
Review/Accept Recommendation ↓ Define Remedial Task	Effect Remedies	
		Test Remedies ↓ Report Results

Figure 15-2. Tenneco improvement process.

formance in the major issue area and to recommend solutions. Recommendations are reviewed and sanctioned by the steering committee, and line management is charged with implementation. The process is iterative; when one issue has been dealt with satisfactorily, another is attacked in the same way. Zent's staff supports the process through facilitation and the provision of useful tools such as statistical evaluations.

Zent, who continues to manage Tenneco's improvement process but now has the title of Executive Director, Productivity and Quality, intends to develop capabilities within the operating units to facilitate the improvement process. With these skills internalized, operating management will not have to rely upon corporate resources, and the process will become ongoing and institutionalized.

The Tenneco approach is noteworthy because of its focus on major business issues and the heavy involvement of senior management in the identification of those issues and in the management of the subsequent improvement process. While employee involvement is less explicit as a driving force for the process than in some other organizations, the participation of employees in addressing the business issues is nonetheless an integral and necessary element of the process.

ETHYL CORPORATION: THE TOTAL FACTOR APPROACH[2]

Ethyl Corporation is a $1.5 billion company headquartered in Richmond, Virginia. Best known for its production of a gasoline additive, Ethyl is actually involved in a variety of businesses, including chemicals, plastics, aluminum, energy, and life insurance.

Like many companies, Ethyl has had a long-standing cost reduction effort in manufacturing based on traditional industrial engineering and process engineering approaches. Improvements typically involved capital investments or manufacturing process changes.

Ethyl's productivity improvement effort broadened considerably in 1981—not because of dissatisfaction with its cost reduction program, but because of the program's success. Millions of dollars of savings had been achieved over the years, and company management recognized that greater opportunities existed through a broadening of the effort.

Initial studies by Ethyl concluded that the American Productivity Center's total factor performance measurement system (Chapter 5 and Appendix B) provided an excellent tool not only for monitoring the productivity of all of the company's resources, but also for evaluating the impact of productivity on profitability. The latter benefit was of particular interest to Ethyl management, for past gains from the cost reduction program did not always correspond to changes in the bottom line. A portion of these gains may well have been passed on to customers, in the form of lower prices, or to employees, in the form of higher wages, but no system existed to track these changes.

The implementation of a total factor measurement system thus became the centerpiece of Ethyl's productivity improvement process. It signaled the organization that productivity improvement was no longer just a manufacturing cost reduction effort, but applied equally to the white collar areas and to all of the company's other input resources as well.

Ethyl management monitors the total factor measurement system closely, not only to evaluate the net effect of all productivity changes, but also to gain insight into the results of pricing actions and to evaluate the impact of product mix changes on the bottom line.

Measurement tools at Ethyl are not limited to the total factor system. The company also utilizes partial productivity ratios to monitor improvement efforts in individual operating units and families of measures (Chapter 5) in white collar operations.

While measurement is clearly the heart of Ethyl's efforts, the company's improvement process is much broader. An organizational infrastructure was established at Ethyl in 1983 with the appointment of L. K. Harmon to the position of Director of Corporate Productivity and the subsequent appointment of a network of part-time productivity coordinators in the various operating divisions and corporate staff departments. In their role as productivity coordinators, these individuals reported to the general manager or department head of their organization. Their efforts were coordinated and supported by Harmon, who reported to the company's Vice President and Budget Director. These relationships are depicted in Figure 15-3.

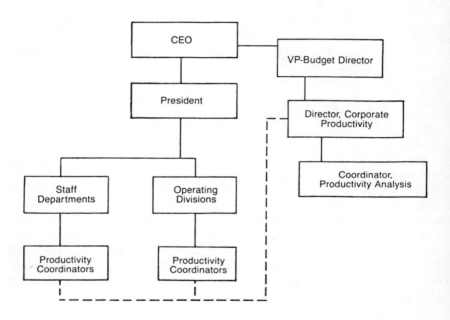

Figure 15-3. Organizational infrastructure for productivity management.

Ethyl has also established a productivity philosophy (Chapter 3) to provide guidance to its managers. The text of this philosophy is presented in Figure 15-4.

Specific improvement activities at Ethyl focus on three strategies:

1. Cost reduction. The company's traditional cost improvement program remains in effect as a viable strategy to improve efficiency and reduce costs. It has been expanded, however, to the white collar and professional areas.

2. Quality and value improvement. Complementing the cost reduction strategy are efforts to focus on quality in manufacturing and on value of services in white collar areas. Statistical process control is a cornerstone of plant quality efforts, while a focus on internal customers supports the service value improvement thrust.

Ethyl Productivity Improvement Program
(A Common Philosophy)

Improving productivity is a way of life throughout the Ethyl Corporation. The Ethyl approach to productivity improvement is based on several key principles that serve as a common philosophy for managers, supervisors, and all employees throughout the entire corporation.

1. The productivity of any operation can be measured by relating its output to the input employed in creating that output.
2. Total productivity refers to a measure of productivity that includes *all* resource inputs and *all* product/service outputs flowing through an operation during a particular operating period.
3. Profitability can be increased by further improving total productivity. Profitability is an economic goal and productivity improvement an operating strategy for achieving that goal.
4. There are two basic implementation strategies associated with improving the productivity of any organization:

   a. Use less input per unit of output (cost reduction), and
   b. Create more output of greater value per unit of input (quality assurance).

5. The task of improving productivity is continuous and unending. There is also no limit to the creativity or the potential contributions of Ethyl employees.
6. Each employee can make some contribution to productivity improvement—all employees should have an opportunity to become involved in the productivity improvement process.
7. The productivity of each element of the Ethyl Corporation can and should be continuously improved—divisions, departments, sections, units, plants, field offices, etc.
8. The management of Ethyl organizations should have an ongoing program of productivity improvement activities that includes at least the following:

   a. Cost reduction
   b. Production/schedule control
   c. Quality/value improvement

Figure 15-4. Ethyl productivity philosophy.

(figure continued on next page)

d. Safe and healthful work environment
e. Employee involvement
f. Coordinated program activities

9. Each division and department manager will benefit from establishing a total productivity improvement program with a designated program coordinator.
10. Program coordination involves encouraging and supporting development of enhanced productivity improvement capabilities and the more effective use of existing capabilities to further improve productivity/profitability.
11. A network of productivity coordinators links each individual coordinator with his counterparts throughout the Ethyl Corporation for information sharing and mutual support.
12. The Ethyl Cost Improvement Program (CIP) is a corporate-wide project/reporting process that supports the cost reduction goals of each organization's total productivity program. Energy conservation is an important part of Ethyl CIP.
13. Ethyl organizations are encouraged to approach quality/value improvement in a manner that is consistent with current customer requirements and local operating capabilities.
14. Ethyl safety and work environment programs are an important element of the corporate-wide total productivity improvement program.
15. Management commitment and involvement is an essential part of the total program.

Figure 15-4. Continued.

3. Improvement in the work environment. Recognizing the key role of the human resource has led Ethyl to its third major thrust: employee involvement. A variety of techniques are utilized to support this avenue, including suggestion systems, quality circles, and productivity committees. In support of the company's total productivity philosophy, employees are encouraged to seek improvement opportunities in all aspects of organizational functioning.

It was suggested earlier that measurement should be kept in perspective as a valuable tool rather than as a prerequisite to improvement or as an end in itself. While a measurement-driven process does carry some risk of overemphasis on measurement, some organizations nonetheless succeed in positioning measurement in a central role without losing sight of the ultimate objective: institutionalizing productivity improvement. Ethyl Corporation appears to be meeting that challenge.

WESTINGHOUSE: THE PRODUCTIVITY AND QUALITY CENTER

The businesses of Westinghouse Electric Corporation are divided into five operating groups:

☐ Energy and Advanced Technology, comprised of high technology businesses in such areas as defense electronics and energy systems.
☐ Industries and International, consisting of businesses associated with the distribution, control, and efficient use of electricity.
☐ Commercial, a diverse collection of specialized businesses, including elevators, transport refrigeration, beverage bottling, and office furniture.
☐ Westinghouse Broadcasting, consisting of radio and television stations, a television production company, and satellite communications.
☐ Westinghouse Credit Corporation, a business finance company.

In the late 1970s, management at Westinghouse became concerned about the impact of foreign competition on many of their markets. It was apparent that the competitive challenge from Japan and other countries had broadened from "low tech" industries, such as steel and textiles, to those that were more technology-driven. Concerned about the threat of foreign challenges to Westinghouse's businesses, company management initiated a high-level task force to examine productivity issues and structure a corporate-wide improvement effort. The work of the task force resulted in the creation in 1980 of a new organizational unit: the Westinghouse Productivity Center.

The Westinghouse Productivity Center brought together under one roof existing corporate staff resources whose primary roles were related to improving productivity and quality. Reflecting its critical importance to the future of the company, the head of the Center is a corporate vice president.

Center management concluded that nothing short of radical cultural change was required if Westinghouse was to thrive in an era of world competition. The traditional management model, which was developed by Frederic Taylor in the early 1900s (Chapter 7), had been fine-tuned for eighty years, but had not changed in its fundamental assumptions: labor should be specialized and is not required to think; quality equates with high cost; inventory has value as a buffer; a certain level of defects is realistic and acceptable; and economic order quantities can be determined through the application of sophisticated formulas.

Key executives at Westinghouse were convinced that a new management model, begun in Japan in the 1960s, was far more effective than the traditional one. Some of the assumptions in this model were that labor specialization is inefficient, that the worker knows his job best, that high quality equates with lowest total cost, that inventories hide problems and are evil, that no defects are acceptable, and that economic order quantities of one should be the goal.

It became apparent to management early in the existence of the Center that the concept of total quality integrated and unified the various elements of the new management model, and the name of the organization was changed to the Westinghouse Productivity and Quality Center (PQC). A definition was adopted as follows: "Total Quality is performance leadership in meeting customer requirements by doing the right things right the first time."

Total Quality is the driving force behind Westinghouse's improvement efforts, applying to every part of the organization and to all employees, over 60 percent of which are white collar workers.

The PQC has adopted three main strategies in support of the company's improvement efforts:

1. Provide the knowledge base to determine operating requirements for total quality.
2. Be a source of technology and systems that meet these requirements.
3. Effect the transfer of these technologies and systems to operating units.

While the PQC encourages each operation to implement Total Quality in its own way, it does define four imperatives which it believes should underlie all company efforts.

☐ **Customer orientation.** The focus of Total Quality is on providing maximum value, relative to price, to an organization's customers, whether internal or external. Performance in this regard is measured by the value/price ratio, a measure promoted by the PQC. Utilizing an internally-developed methodology to quantify value, this measure provides an indicator of relative value, as defined by the customer.

☐ **Human resource excellence.** Recognizing that people are the key to success, the Westinghouse effort emphasizes education, training, and involvement in order to enhance employee contributions to quality and productivity improvement. Among the PQC's training initiatives is the Westinghouse Quality College, and Westinghouse has long been one of the nation's leading users of quality circles.

☐ **Product/process leadership.** Intent on achieving world-class, error-free performance, the PQC promotes a systems approach which stresses that products and services must be jointly designed with the processes that produce them. This integration of product and process design minimizes waste and maximizes the value received by the customer relative to the cost of providing the product or service.

☐ **Management leadership.** Management's commitment to Total Quality and its leadership in establishing Total Quality as a way of life is the fourth imperative. Emphasis is placed on setting improvement objectives, communicating clearly and consistently about the initiative, providing appropriate incentives and motivation, and effectively utilizing measurement as a feedback tool.

The PQC provides a variety of tools and services to operating units in support of the Total Quality effort. One of these is the Total Quality Fitness Review, an assessment process designed to evaluate an operation's programs and practices in order to identify strengths and weaknesses, and provide recommendations for improvement. This assessment methodology is organized around the four imperatives, which are further subdivided into 12 "conditions of excellence." This model, as graphically depicted by the PQC, is shown in Figure 15-5.

Improvement tools provided by the PQC include OPTIM (Operating Profit Through Time and Investment Management), a structured analysis of the buildup of cost over time as a product or service passes through its production cycle; and O:TIME (Organization: Techniques for Improving Managed Cost-Effectiveness), a systematic analysis of an organization's tasks and functions for the purpose of reducing costs and cycle time and improving quality.

With its 140 employees and a "board of directors" consisting of top business unit managers, the Westinghouse Productivity and Quality Center represents a major management commitment to organizational change.

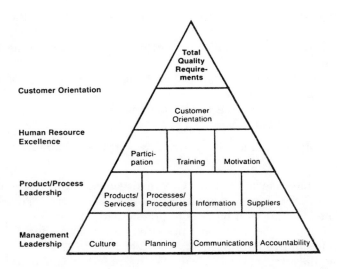

Conditions of Excellence to Fulfill Total Quality Requirements

Figure 15-5. Total quality fitness review.

CITY OF PHOENIX: PRODUCTIVITY IN THE PUBLIC SECTOR

While the public sector in general has lagged behind industry in adopting productivity improvement programs, the City of Phoenix is an exception, with an effort that dates back to 1970.[3,4]

The force behind the City's effort was the burgeoning population growth that began in the 1960s. This growth resulted in dramatic increases in demand for the city's services. Since revenue growth did not keep pace with this demand, productivity became a major concern.

Following the recommendations of a consultant, the City established in 1970 a Work Planning and Control Program. Basically an industrial engineering program, this effort focused on work measurement and methods improvement. Resources were brought to bear in the form of an operations analysis division, which was created to assist city departments in implementing these activities.

While this initial program did result in some significant improvements, maintenance of the work standards became cumbersome, and an inordinate focus on the numbers caused some dysfunctional behaviors to occur. Accordingly, the effort was redirected in 1978.

The new direction provided for greater focus on the human resource. This focus is apparent in the goals that were established for the new plan:

☐ To create an atmosphere of cooperation and coordination among all city employees.
☐ To harmonize the goals of the individual and the goals of the organization.
☐ To develop a positive, challenging attitude toward existing methods and procedures.
☐ To develop skills for effectively identifying and evaluating opportunities for productivity improvement.
☐ To improve city operations to provide reduced costs as well as to improve the quality of work life.

The revitalized productivity program was structured around ten key elements:

1. Top Management Support. City management meets frequently to explore improvement opportunities, task forces are regularly established to deal with barriers to productivity improvement, and compensation and goal-setting practices have been modified to support and reinforce improvement.
2. Organizational Development. The use of industrial engineering techniques has been modified to emphasize human factors and to incorporate the principles of organizational change.

3. Management by Objectives. Departmental objectives and productivity improvement goals are established, with compensation increases tied to goal achievement. Organizational climate objectives are included along with more traditional performance objectives.
4. Employee Suggestion Program. While actually in existence since 1950, the suggestion system was provided increased emphasis, with noteworthy results: a tenfold increase in employee participation and dramatic increases in documented savings and cash awards.
5. Productivity Training. Employee training activities were substantially increased, covering such topics as work sampling techniques, forms and records management, time management, flow process charting, work distribution analysis, and project management.
6. Technology Transfer. Technology sharing networks were established with other cities in order to keep abreast of new developments in urban delivery systems.
7. Employee Skills Training. Numerous in-house training programs and a tuition reimbursement plan were developed in order to increase employee skills and abilities.
8. Citizens' Productivity Advisory Committee. A committee representing a variety of community interests reviews the city's productivity program and recommends changes.
9. Productivity Organization. To ensure that productivity improvement remains visible, a productivity coordinator was appointed for each city department.
10. Central Staff Assistance. Support to line departments is provided through operational studies and guidance provided by a centralized support group.

The last point deserves some elaboration. The original operations analysis staff was combined with a training and development staff in 1982 to form the Value Management Resource Office. The merger of these two groups produced an unusual combination of industrial engineering and human resource expertise and signalled the integration of these disciplines in the pursuit of productivity improvement.

The Value Management Resource Office provides a full range of consulting services and training programs to city departments. Consulting services are divided into three categories, as summarized here:

Diagnostic Services

- Employee and/or client surveys
- Work load distribution
- Problem diagnosis
- Work output measures

Behavioral Services

- Team building
- Work group development
- Management consultation
- Conflict resolution
- Customized training

Analytical Services

- Work measurement/standards
- Methods improvement
- Work flow and scheduling
- Office systems and procedures
- Productivity training
- Decision making assistance

The City of Phoenix is a pioneer in public sector productivity improvement and provides an excellent illustration of a productivity effort that has evolved from a traditional, industrial engineering approach to one that recognizes that people are the key to success.

XEROX: PRODUCTIVITY THROUGH PEOPLE

Xerox Corporation, the large, multinational manufacturer of office equipment, with over 100,000 employees worldwide, is a premier example of a large organization that is undertaking a human-resource based change effort to reshape its culture in order to survive in an era of heightened world competition.

In 1970, Xerox was essentially the sole manufacturer of plain paper reprographic products. As the seventies unfolded, however, strong competitors such as IBM, Kodak, and Canon invaded Xerox's markets with new technologies, and the company rapidly lost market share. Before the decade was out, the management of Xerox recognized that the company was no longer competitive. The company's survival was at stake.

In response to that threat, a task force during 1979–80 conducted research into human resource issues and their impact on strategic change. An outcome of this study was the initiation by the company of a quality of work life/employee involvement (QWL/EI) effort in the North American Manufacturing Division (NAMD).

It was apparent from the research conducted that the commitment of the union was vital to the success of the QWL/EI effort. Management accordingly brought the proposed process to the bargaining table in late 1979. The leadership of the Amalgamated Clothing and Textile Workers Union recognized the gravity of the company's competitive position, and the contract that emerged from the bargaining process in early 1980 sanctioned a joint labor-management effort. A policy statement governing the QWL/EI process was drawn and endorsed by both management and the union.

The joint nature of the improvement process was reflected in the structures established to oversee the QWL/EI effort. A joint planning/policy committee had overall responsibility for launching and governing the process. The membership of this committee included high-ranking officers of the ACTWU and the International Union of Operating Engineers, as well as senior management. In addition, a plant advisory committee was established at each plant to assist in implementing the process at the local level. This committee consisted of local union officials, union members, the plant manager, and other management representatives. At still another level down were business center steering committees, whose function was to organize and support problem-solving teams in various areas of the plant. Finally, coordinator/trainers were appointed from both the union and management ranks to train and assist the problem-solving teams.

The QWL/EI effort was integrated with the company's competitive benchmarking process, which was begun as a separate initiative in 1979. Defined by the company as "the continuous process of measuring our products, services, and practices against our toughest competitors or those companies renowned as the leaders," competitive benchmarking served as a motivating vehicle by creating a felt need among employees to improve organizational performance.

From its beginnings in NAMD, employee involvement has grown into a company-wide change effort, overseen by Dominick R. Argona, Manager, Employee Involvement Program Development. The change process is guided by a three-phase strategy requiring over a decade to implement. The strategy is presented graphically in Figure 15-6.

The first stage of the strategy—QWL and Employee Involvement—is characterized by the use of group problem-solving structures. The particular type of structure utilized at a given location is not dictated by the strategy; a variety of options are available, including task forces, quality circles, study teams (groups of employees relieved of their normal job responsibilities for up to six months in order to intensively study a work area to make it more

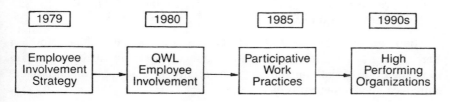

Figure 15-6. Evolution of employee involvement at Xerox.

competitive), and planning teams (groups of employees who examine strategic and operational issues and make recommendations for the future). The selection of an appropriate structure is a local decision, and the effort is managed jointly at all locations by labor and management.

The second phase of the strategy, Participative Work Practices, involves the application of sociotechnical design principles to redesign work organizations. The focus of this second generation of employee involvement is the business area work group (BAWG), a cross-functional team of employees, typically 25–30 in number, that is formed around a specific product or service. The BAWGs are intended to be managed as small, entrepreneurial businesses linked to other parts of the organization through explicit customer/supplier relationships. Here again, a variety of design options are possible, but the goal is to achieve the highest degree of self-management possible.

The third phase of the strategy is labeled High Performing Organizations. In this phase, the focus is on maximizing the organization's ability to adapt to change through the development of shared goals, environmental sensing processes, union involvement in business planning, maximum teamwork, and self-managing work groups.

In 1983, Xerox launched another organization-wide initiative called Leadership Through Quality. It too is a long-term change process designed to create an environment of constant improvement in meeting customer requirements. In a pamphlet issued to its employees, the company defined how Leadership Through Quality differs from the traditional approaches:

□ Whereas the conventional definition of quality reminds us of words like "goodness" and "luxury," Xerox defines quality as "conformance to customer requirements."

□ Whereas the conventional performance standard for quality is some acceptable level of defects or errors, the quality performance standard at Xerox is "products and services that fully satisfy the requirements of our customers."

□ Whereas the conventional system of achieving quality is to detect and correct products after they have been completed, Xerox emphasizes the "prevention of errors."

□ Whereas the conventional system for measurement of quality relies on indices, Xerox measures quality by the "cost we incur when we do not satisfy customer requirements."

The Leadership Through Quality process is not a stand-alone effort for Xerox, but has been integrated with employee involvement to build upon the foundation of skills and employee commitment that has been laid by the employee involvement process.

Xerox provides one of the leading examples of a company that is committed to improving productivity through people. It has invested enormous amounts of resources in a long-term effort to fundamentally change the way in which it manages its human resources.

REFERENCES

1. Belcher, John G., Jr., *Manager's Notebook: Productivity Focus: A Structure for Organizational Diagnosis and Planning.* Houston, TX: American Productivity Center, 1984.
2. *Case Study 54: Ethyl Corporation.* Houston, TX: American Productivity Center, 1986.
3. *Case Study 38: City of Phoenix.* Houston, TX: American Productivity Center, 1984.
4. Aranda, Eileen Kelly, "Public Sector Productivity: A Focus on Phoenix," *National Productivity Review,* Summer, 1982.

16

The Productivity Perspective

SUMMING UP

The preceding chapters have suggested that if an organization is to be successful in achieving meaningful, continuing productivity improvement, it must raise productivity to the level of a strategic issue and view productivity improvement from the perspective of a management process—an integral, institutionalized element of organizational functioning.

This perspective on productivity has significant implications for the organization; a productivity "program" is no longer an acceptable response to the productivity challenge. A major organizational change effort must be undertaken, and the objective of this initiative must be nothing less than a change in organizational culture. Productivity must become a part of the very fiber of the organization, supported and reinforced by the other processes, systems, and practices through which the organization is managed on a day-to-day basis.

PROFILE OF A HIGH-PERFORMING ORGANIZATION

An organization that has an effective productivity management process will have certain common features regardless of its industry or the nature of its business:

☐ *Management commitment* to productivity improvement will be evident and unambiguous. Managers at all levels will be actively supporting productivity improvement through ongoing communications, allocation of resources, and personal involvement.
☐ *Organizational awareness* of productivity will be high. All employees will understand productivity and its implications, and the work force will be

continually informed about competitive challenges and organizational progress in improving productivity.

☐ *Management responsibilities* for productivity improvement will be explicit, and accountabilities will be strong. All managers will view productivity improvement as a key element of their job responsibilities, and productivity will be reinforced by the performance appraisal and promotional processes.

☐ *Productivity goals* will exist at all organizational levels, and progress against these goals will be monitored closely and publicized.

☐ The *reward system*, both financial and nonfinancial, will strongly reinforce productivity improvement. Recognition for exceptional performance will be prevalent, and employees will share financially in the fruits of organizational productivity improvement.

☐ *Productivity measures* will be widespread and visible, and they will be used for feedback, recognition, opportunity assessment, and problem-solving.

☐ *Quality* improvement will be pursued as a key performance variable and will be viewed as a means to productivity improvement.

☐ *Employee involvement* in productivity improvement will be an organizational norm, supported by management at all levels. The human resource environment will be a constant concern of management, as people will be viewed as the key to productivity improvement.

☐ The productivity of *all resources*—capital, materials, and energy, as well as labor—will be addressed through the productivity improvement process.

☐ *Improvement techniques* will be widely utilized, but their use will be tailored to the circumstances of each organizational unit and they will be viewed as tools rather than as ends in themselves.

☐ Productivity will be an integral element of the *budgeting, planning and financial reporting systems*. Productivity will be treated explicitly and prominently in these systems.

The absence of any of these elements renders the productivity management process incomplete and reduces the likelihood of achieving a productivity-driven culture.

A FINAL THOUGHT

In the end, management must decide whether it wants a productivity program or a productivity management process. If it does not make a conscious decision on this issue, a program will result by default. If management does not have the energy nor the inclination to undertake a long-term effort to develop a productivity management process, they should at least reach that conclusion explicitly and adjust their expectations accordingly.

Management should be aware, however, that the ramifications of the productivity challenge have never been greater. The failure of American management to maintain its historical productivity growth rate over the last 20 years has produced incalculable damage. Countless jobs have been lost to foreign producers and entire industries have been laid low. If we continue down this path, our free enterprise system itself will surely be threatened.

The old, traditional approaches to managing people and productivity are wholly inadequate in this era of rapid change and intense competition.

The productivity challenge demands an urgent and dramatic response.

APPENDIX A

Productivity Executive Job Descriptions

Following are a collection of job descriptions for productivity executives. These documents are in actual use in organizations and encompass a variety of industries.

POSITION DESCRIPTION: DIRECTOR—PRODUCTIVITY SERVICES

Concept and Purpose

Productivity growth is essential for maintaining the American economic system and improving our standard of living. It is a vital underpinning to the private enterprise system and one of its chief hallmarks. The company recognizes the importance of productivity improvement and the role it must play if the company is to grow, remain competitive, and continue to be profitable.

The purpose of Productivity Services is to (1) foster and maintain an awareness of the need for gains in productivity and quality of working life throughout the company, (2) develop techniques for improvement, and (3) assist divisions and units in achieving improvements.

Nature and Scope

This program has been established by the Chairman of the Board as a high-priority effort to raise the company's productivity measure annually. This goal has been assigned to the division general managers and they have incorporated it into their annual operating goals. The incumbent works to assist managers across the country to meet this demanding objective.

Corporate Productivity Services operates within general company policies and an approved operating plan. It is nonduplicative of efforts of operating units. Its services and activities are performed based on the following criteria:

1. Cost effectiveness
2. Need for corporate continuity
3. Time constraints
4. Scope of tasks

Principal Accountabilities

1. Initiating and maintaining effective productivity awareness and communications programs.
2. Serving as an advisor and consultant to operating units regarding their individual productivity programs.
3. Monitoring the key factors contributing to low productivity and developing and recommending corrective actions.
4. Maintaining liaison with the American Productivity Center.
5. Performing data collection and analysis relative to national, international, and industry trends, inter-firm and intra-firm comparisons, and highlighting potential problems.
6. Ensuring that appropriate education and training in productivity improvement are accomplished throughout the company.
7. Identifying, compiling, and cataloging a repertoire of productivity improvement techniques which can be referred to and drawn upon by the operating units.
8. Supporting the objectives of the other components of the Corporate Staff.

Experience

The above accountabilities of the Director of Productivity Services requires an individual experienced both in line and staff positions.

This position requires that the incumbent possess specialized experience in the development and management of major programs having company-wide impact.

A qualified individual should possess outstanding communication and interpersonal skills, a good knowledge of the free-enterprise system, and a broad business background. The incumbent works as an individual contributor, but must work well with a wide variety of managers to coordinate the overall productivity improvement effort throughout the company.

Education

Qualified candidates for this position should possess an advanced degree in business. Expertise in the area of the behavioral sciences is desirable.

TITLE: CORPORATE PRODUCTIVITY CONSULTANT

Brief Description of Function

Act as a consultant in productivity improvement to the president, head of groups, divisions, and corporate staff functions and other organizations. Develop programs to exploit opportunities/solve problems in productivity. Provide impetus and direction to the corporate productivity improvement strategy.

Specific Activities

A. Be a communicator/catalyst to gather and disseminate productivity improvement ideas.

- The consultant will gather productivity improvement ideas from within the company and other sources (APC, professional societies, trade associations, educational institutions and consultants).
- He will become an active member of each group's/division's productivity task force to aid in spreading this information.
- There are a number of consultants working in the company; becoming familiar with their services, capabilities, and costs is a part of this function.

B. Design and/or aid in designing workshops, seminars, conferences, and programs—coordinating the efforts with our staff groups.

- The consultant will work with Human Resources on training for productivity improvement.
- He will cooperate with corporate communications on a number of messages—awareness of the problem, significant successes both inside and outside of the company, what the company is doing in productivity, etc.
- Recommending outside programs and seminars will be a part of the consultant's function.

C. Be a member of the Corporate Productivity Task Force and a source of programs of interest to it.
D. Organize and/or work with special project task forces in divisions/ groups to work on specific aspects of productivity improvement.
E. Work with any officer of the corporation who has a special need for productivity expertise.

POSITION DESCRIPTION: PRODUCTIVITY IMPROVEMENT PROGRAM MANAGER

Reports to: Operations Director
Supervises: Staff, as required.
Basic Function: Establishes, manages, coordinates, develops, implements and maintains a productivity improvement program within the Division. The position places heavy emphasis on advanced manufacturing technology and planning necessary to support the Division's needs for facilities, tooling, design, manufacturing methodology and materials to maximize the Division's resources, and opportunities both near and long-term; recommends, develops, and implements plans and procedures to organize and staff for optimum performance and execution of the advanced technological manufacturing methods and planning process; also coordinates with vendors and customers as required to carry out his responsibilities.

Major Duties and Responsibilities

- Establish and develop this Divisional activity involving personnel, existing and planned facilities, equipment, manufacturing methods, machinery systems, metalworking, and computer-aided manufacturing and design.
- Be responsible for the investigation of, planning for, design, development and implementation of new equipment for the manufacturing of new products.
- Ensure that the type of equipment, tooling, fixtures, etc. satisfy product and quality engineering specifications for the new products being manufactured.
- Supervise the installation of new equipment or major changes in existing equipment intended for the production of new products.
- Serve as the Division's focal point in computer-aided manufacturing and design.
- Keep fully up to date with the evolution of metalworking, seeking constantly to find new applications which will materially benefit the Division.
- Develop and help implement strategic manufacturing plans to ensure that the necessary processes, equipment, and facilities are available at the proper time and place to achieve their projected goals.
- Coordinate the introduction of a new product or product change, equipment to be used in the manufacturing and assembly process, the process of materials to produce that product, and the labor controls and standards applied.
- Evaluate the Division's capabilities to meet production requirements to ensure that future requirements are met.

- Recommend Division productivity policies for overall objectives to the Division Vice President and General Manager, and carry out those which are approved.
- Develop and evaluate the overall effectiveness of the division productivity plans and objectives.
- Develop standard productivity performance reporting techniques and formats.
- Prepare Division productivity procedures and working relationships to assure optimum utilization of the technical resources, especially personnel, in providing for a logical flow of products, as required, and see to their implementation.
- By close association with production operations, engineering and other technical functions, assure continuing attention is given to improving manufacturing methods, developing new equipment, improving plant and area layouts, and upgrading operating procedures.
- Monitor accomplishments against established goals.
- Develop and implement training process required for productivity improvement.
- Maintain a productive, quality-conscious, harmonious work force.
- Work with both suppliers and customers as this is required to meet the needs of activities within his realm of responsibilities.
- Provide a trained and motivated staff to assure continuity and achievement of his function wherever it is called upon in the Division.
- Manage his function within the limits established by capital and expense budget plans approved by his superior.
- Keep his supervisor informed on all pertinent projects, programs, and activities.

APPENDIX B

The APC Performance Measurement System

The total performance measurement system described in Chapter 5 was developed by Carl Thor of the American Productivity Center based on work initially done by the National Productivity Institute of the Republic of South Africa. The system provides a means of analyzing profitability in terms of its two major components: productivity and price recovery. It thus provides the connecting link between productivity analysis and the income statement.

Data are provided to the system in terms of the following equation:

Value = Quantity × Price

For every output and every input of an organization, the dollar value appearing on the income statement can be viewed as consisting of a physical quantity multiplied by its unit price or unit cost. The financial value of the labor input, for example, is derived by multiplying the quantity of man-hours worked by the average hourly wage rate.

When all outputs and all inputs are expressed in the $V = Q \times P$ format, the relative changes, over time, between output values, quantities, and prices and input values, quantities, and prices can be calculated to provide the desired information (Figure B-1).

Figure B-1. Relationship between output and input—values, quantity, price.

The comparison between the relative changes in the value of the outputs and the value of the inputs yields the change in profitability. The relationship between quantities of output and quantities of input over time is productivity. This is consistent with the notion of productivity as the relationship between physical outputs and physical inputs. Finally, comparing changes in unit selling prices to unit input costs yields price recovery, or the ability of the organization to pass through its unit cost changes in selling prices.

The basic data for a hypothetical manufacturer of chairs and tables are presented in Table B-1. Only two of the three requisite data elements (value,

Table B-1
Basic Data for Performance Measurement System

	Period 1			Period 2		
	Value	Quantity	Price	Value	Quantity	Price
Output						
Chairs	50,000	1,000	50.00	66,000	1,200	55.00
Tables	40,000	200	200.00	33,600	160	210.00
Total output	90,000			99,600		
Input						
Materials						
Maple stock	20,000	20,000	1.00	25,200	21,000	1.20
Varnish	1,000	100	10.00	1,200	100	12.00
Screws	200	200	1.00	160	148	1.08
Total materials	21,200			26,560		
Labor						
Woodworker	24,000	4,000	6.00	30,400	3,800	8.00
Finisher	8,000	1,000	8.00	8,320	800	10.40
Total labor	32,000			38,720		
Energy						
Electricity	3,000	30,000	0.10	3,780	27,000	0.14
Capital						
Cash	600	8,000	0.075	560	7,000	0.080
Leases	1,800	24,000	0.075	1,920	24,000	0.080
Inventory	900	12,000	0.075	810	10,125	0.080
Depreciation	15,000	300,000	0.050	15,300	300,000	0.051
Pretax return	14,100	300,000	0.047	15,120	315,000*	0.048
Total capital	32,400			33,710		
Miscellaneous						
Taxes & insurance	1,400	1,000	1.40	1,500	1,000	1.50
Total input	90,000			104,270		
Difference	0			(4,670)		

* Added land was purchased for $15,300 at beginning of Period 2 and is deflated to Period 1 price levels.

quantity, and price) are required for each output and input, as the third is then calculable. These data can generally be derived from an organization's accounting and information systems.

The major exception is in the capital section, where certain approaches that are not consistent with conventional accounting practices must be adopted:

☐ Lease expense is included as a capital input. The organization must be concerned with the effectiveness with which it utilizes all of its physical plant and equipment, whether owned or leased. The Quantity column represents the capitalized value of the lease.

☐ A Return component is included as an input. The Return represents the payment required by the owners of capital to supply that capital to the organization. Viewed in another way, the Return represents the opportunity cost associated with the commitment of capital to the organization as opposed to an alternative use.

☐ Depreciation, representing the annualized input of physical capital, differs from accounting depreciation. Since the Quantity column requires a consistent representation of physical input over time, the impact of inflation must be removed from the organization's fixed asset base before depreciation is calculated. The proper treatment, then, is to calculate Quantity by restating fixed assets at replacement cost (or any base-year cost) and then applying an economic depreciation rate (Price) to obtain the Value.

☐ Complications arise in Period 2, as physical equivalencies must be maintained in the Quantity column relative to Period 1; this requires the deflation of financial capital elements (cash and inventories) to Period 1 price levels. In addition, the numbers in the Price column for Depreciation and Return must be incremented for inflation; otherwise, the calculated Value for these elements will be insufficient in an inflationary environment.

Because of the various conventions described above, and because output would normally be based on sales value of production rather than sales, some reconciling items are necessary in order to equate the Value column with the organization's income statement.

Once the basic data are organized as shown in Table B-1, two sets of calculations are required to derive performance indexes. In the first calculation, each number in Period 2 is divided by its counterpart in Period 1. The results, called Change Ratios (Table B-2), represent the relative change in the value, quantity and price of each output and input from one period to the next.

Table B-2
Change Ratios

	V2/V1	Q2/Q1	P2/P1
Output			
Chairs	1.3200	1.2000	1.1000
Tables	0.8400	0.8000	1.0500
Total output	1.1067	1.0222*	1.0826*
Input			
Materials			
Maple stock	1.2600	1.0500	1.2000
Varnish	1.2000	1.0000	1.2000
Screws	0.8000	0.7400	1.0800
Total materials	1.2528	1.0447*	1.1992*
Labor			
Woodworker	1.2667	0.9500	1.3333
Finisher	1.0400	0.8000	1.3000
Total labor	1.2100	0.9125*	1.3260*
Energy			
Electricity	1.2600	0.9000	1.4000
Capital			
Cash	0.9333	0.8750	1.0667
Leases	1.0667	1.0000	1.0667
Inventory	0.9000	0.8438	1.0667
Depreciation	1.0200	1.0000	1.0200
Pretax return	1.0723	1.0500	1.0213
Total capital	1.0404	1.0151*	1.0249*
Miscellaneous			
Taxes & insurance	1.0714	1.0000	1.0714
Total input	1.1586	0.9815*	1.1804*

* Weighted

For use in later calculations, it is also necessary to calculate Change Ratios for each Total (Total output, Total materials, etc.). This does not present a problem in the Value column, as totals are available in the raw data. For the Quantity and Price columns, however, a weighting scheme is required in order to develop "totals" that can then be divided to produce Change Ratios. The recommended approach is to use base-period price-weighting (multiply each element's quantity in each period by the element's Period 1 price) to develop Quantity totals and current-period quantity-weighting (multiply each element's price in each period by the element's Period 2 quantity) to develop Price totals. Change Ratios for each Total line can then be calculated as described in the preceding paragraph. The symmetrically opposite weighting approaches are suggested for obscure, but technically necessary, reasons.

In the second calculation, the Change Ratio for each input is divided into the corresponding Change Ratio for Total Output. The resulting Performance Ratios provide indexes of profitability, productivity, and price recovery (Table B-3). Thus, the Productivity Performance Ratios relate the change in the quantity of total output to the change in quantity of each input. The Performance Ratios on the Total Input line provide measures of total productivity and total price recovery. Now the organization can objectively determine whether it is a Company A or a Company B (Chapter 1).

Additional analytical data can be derived from this system by calculating the dollar impact on profits of each of the changes indicated by the Performance Ratios. This is accomplished in the Profitability and the Productivity columns by subtracting each Input Change Ratio from the Total Output Change Ratio and multiplying the resulting number by that input's Value in

Table B-3
Performance Ratios and Effect on Profits

	Performance Ratios			Effect on Profits		
	Profit-ability	Produc-tivity	Price Recovery	Profit-ability	Produc-tivity	Price Recovery
Output						
Chairs						
Tables						
Total output						
Input						
Materials						
Maple stock	0.8783	0.9735	0.9022	(3,067)	(556)	(2,511)
Varnish	0.9222	1.0222	0.9022	(93)	22	(115)
Screws	1.3834	1.3814	1.0014	61	56	5
Total materials	0.8834	0.9785	0.9028	(3,099)	(478)	(2,621)
Labor						
Woodworker	0.8737	1.0760	0.8120	(3,840)	1,733	(5,573)
Finisher	1.0641	1.2778	0.8328	533	1,778	(1,245)
Total labor	0.9146	1.1202	0.8165	(3,307)	3,511	(6,818)
Energy						
Electricity	0.8783	1.1358	0.7733	(460)	367	(827)
Capital						
Cash	1.1857	1.1682	1.0150	104	88	16
Leases	1.1429	1.0222	1.1180	72	40	32
Inventory	1.2296	1.2115	1.0150	186	161	25
Depreciation	1.0850	1.0222	1.0614	1,300	333	967
Pretax return	1.0320	0.9735	1.0601	485	(392)	877
Total capital	1.0637	0.9878	1.0768	2,147	230	1,917
Miscellaneous						
Taxes & insurance	1.0329	1.0222	1.0105	49	31	18
Total input	0.9552	1.0415	0.9171	(4,670)	3,661	(8,331)

Period 1. For the Price Recovery column, the effect on profits should be backed into by subtracting the productivity effect from the profitability effect. These data are also displayed in Table B-3.

The Effect on Profits number shown on the Total Input line corresponds to the change in reported profits from Period 1 to Period 2 (after the reconciling adjustments described above).

The American Productivity Center's system can treat any two sets of data; it can be used, for example, to analyze actual results versus budget as well as year-to-year changes.

APPENDIX C

Readiness Assessment Questionnaire

The characteristics on the following pages may reflect your organization's readiness to support a successful employee involvement process. These characteristics may also indicate the amount of organizational preparation which will be needed before your employee involvement effort can be implemented.

Consider each of the following statements carefully, and if you feel that you have the necessary information, indicate the extent to which you can agree that the statement fits your perceptions of your organization by circling your responses. If you have no opinion or incomplete information to form an opinion, circle the "Don't Know" category.

In assessing your organization you should consider both the specific work unit which is targeted for an employee involvement program *and* the broad overall organizational context in which this employee involvement effort is being undertaken. If there is a difference between the characteristics of the specific work unit and the overall organization, you may want to note this in your response by coding separate responses for the two groups.

Organizational Characteristics

In my organization:	Strongly Agree	Agree	Neither Agree nor Disagree	Disagree	Strongly Disagree	Don't Know
1. . . . people can be rewarded and recognized for teamwork accomplishments.	SA	A	N	D	SD	?
2. . . . measurement data exist to describe important performance results.	SA	A	N	D	SD	?
3. . . . measurement data are regularly shared with employees to let them know how they are doing.	SA	A	N	D	SD	?
4. . . . the labor-management climate is more cooperative than adversarial.	SA	A	N	D	SD	?
5. . . . communications methods exist which allow for a 2-way flow of information between management & employees.	SA	A	N	D	SD	?
6. . . . departments cooperate to achieve common goals.	SA	A	N	D	SD	?

(questionnaire continued on next page)

Organizational Characteristics (continued)

	Strongly Agree	Agree	Neither Agree nor Disagree	Disagree	Strongly Disagree	Don't Know
7. . . . personnel policies & practices are based on the assumption that employees want to do a good job.	SA	A	N	D	SD	?
8. . . . time and money are spent on the training and development of people.	SA	A	N	D	SD	?

Employee Characteristics

Most of the employees in my organization:

	Strongly Agree	Agree	Neither Agree nor Disagree	Disagree	Strongly Disagree	Don't Know
9. . . . understand the need for productivity/quality improvement.	SA	A	N	D	SD	?
10. . . . understand how their jobs contribute to team efforts and organizational success.	SA	A	N	D	SD	?
11. . . . feel comfortable in sharing their ideas & concerns with their immediate supervisor.	SA	A	N	D	SD	?
12. . . . believe that management is concerned with their needs as well as the organization's needs.	SA	A	N	D	SD	?

Supervisor Characteristics

Most of the first-line supervisors in my organization:

	Strongly Agree	Agree	Neither Agree nor Disagree	Disagree	Strongly Disagree	Don't Know
13. . . . perceive their role primarily as teachers, developers, & coaches.	SA	A	N	D	SD	?
14. . . . are not threatened by the knowledge, skills & contributions of *their subordinates*.	SA	A	N	D	SD	?
15. . . . are experienced & skilled in leading meetings which encourage participation.	SA	A	N	D	SD	?
16. . . . believe that they will be rewarded for successfully adopting a participative management style.	SA	A	N	D	SD	?
17. . . . encourage employees to engage in department problem solving when appropriate.	SA	A	N	D	SD	?

Management Characteristics

Most managers and supervisors in my organization:

	Strongly Agree	Agree	Neither Agree nor Disagree	Disagree	Strongly Disagree	Don't Know
18. . . . believe that employee participation can further the organization's goals as well as the individual employee's goals.	SA	A	N	D	SD	?
19. . . . express respect for employees as valuable contributors to the organization.	SA	A	N	D	SD	?
20. . . . express concern for employees' well-being & job satisfaction.	SA	A	N	D	SD	?
21. . . . frequently express appreciation for the contributions which employees make.	SA	A	N	D	SD	?

22. . . . routinely share business information with employees.	SA	A	N	D	SD	?
23. . . . explain organizational procedures & policies to employees.	SA	A	N	D	SD	?
24. . . . are receptive to employees' inputs to the decision-making process.	SA	A	N	D	SD	?
25. . . . listen to suggestions and seriously consider them.	SA	A	N	D	SD	?
26. . . . do a good job of coaching & training people to help them improve their performance.	SA	A	N	D	SD	?
27. . . . expect employees to use initiative. If I am confident and I have the right approach, I am expected to act on it.	SA	A	N	D	SD	?

Labor-Management Characteristics

	Strongly Agree	Agree	Neither Agree nor Disagree	Disagree	Strongly Disagree	Don't Know
Management in my organization:						
28. . . . routinely shares business information with the union.	SA	A	N	D	SD	?
29. . . . solicits union views and ideas regarding opportunities for improvement.	SA	A	N	D	SD	?
30. . . . acknowledges union contributions to organizational accomplishments when appropriate.	SA	A	N	D	SD	?
The union is:						
31. . . . interested in exploring different ways of carrying out the Labor-Management relationship.	SA	A	N	D	SD	?

INDEX